The **Sustainable Development Goals Series** is Springer Nature's inaugural cross-imprint book series that addresses and supports the United Nations' seventeen Sustainable Development Goals. The series fosters comprehensive research focused on these global targets and endeavours to address some of society's greatest grand challenges. The SDGs are inherently multidisciplinary, and they bring people working across different fields together and working towards a common goal. In this spirit, the Sustainable Development Goals series is the first at Springer Nature to publish books under both the Springer and Palgrave Macmillan imprints, bringing the strengths of our imprints together.

The Sustainable Development Goals Series is organized into eighteen subseries: one subseries based around each of the seventeen respective Sustainable Development Goals, and an eighteenth subseries, "Connecting the Goals", which serves as a home for volumes addressing multiple goals or studying the SDGs as a whole. Each subseries is guided by an expert Subseries Advisor with years or decades of experience studying and addressing core components of their respective Goal.

The SDG Series has a remit as broad as the SDGs themselves, and contributions are welcome from scientists, academics, policymakers, and researchers working in fields related to any of the seventeen goals. If you are interested in contributing a monograph or curated volume to the series, please contact the Publishers: Zachary Romano [Springer; zachary.romano@springer.com] and Rachael Ballard [Palgrave Macmillan; rachael.ballard@palgrave.com].

Vinesh Chandra · Margaret Lloyd ·
Dann Mallet
Editors

Integrated STEM Education in Papua New Guinea

Teacher Professional Learning and Praxis

Springer

there were abrupt stops to delivering short courses traditionally offered on university campuses. The design of the Graduate Certificate in STEM Education course was such that it could withstand the shocks that were associated with the rapidly changing COVID mandates. Because of this flexibility, the QUT team, in partnership with Australia Awards and PNG's National Department of Education, devised innovative and dynamic approaches to overcome COVID-19 border closure challenges so that the academic progress of our teachers enrolled in the course was not hampered regardless of whether they were based in urban, rural, or remote locations. In addition, many other challenges akin to developing countries were tackled so that customised delivery models could be enacted effectively and efficiently.

For practitioners in PNG and elsewhere, this book demonstrates the impact of a well-designed and delivered postgraduate course on teachers' practice. Teachers seldom get opportunities to tell their stories in this way. Each chapter links to one of the four discrete contexts of STEM education: Authentic Learning, Design Thinking, Project-Based Learning, and Teacher Leadership. This book has created a unique opportunity for the PNG teachers to immerse themselves as authors and outline how the projects they initiated address their classroom and school needs. With reviews by university academics, teachers' narratives present contextualised insights into how an integrated approach to STEM Education can be adopted and implemented. The alignment of the teachers' chapters with the United Nation's Sustainable Development Goals (SDGs) demonstrates how teaching and learning at the grassroots level can add value to significant global challenges and initiatives. It enables the students to think more deeply about why they are learning what they are learning.

This book adds new knowledge on how postgraduate training in integrated STEM Education can be designed and delivered to in-service high school teachers in PNG, other countries in the Indo-Pacific, and elsewhere. All such courses should be contextually appropriate so that participating teachers can easily apply their knowledge and skills within their contexts. Allowing teachers to explain their initiatives through book chapters and other publications facilitates knowledge sharing. In this way, the value of teachers' professional learning extends beyond their classrooms and schools.

Dr. Uke Kombra PhD, OBE
Secretary for Education (Papua New Guinea)

Port Moresby, Papua New Guinea

Acknowledgements

We acknowledge the First Nations people of the lands where this book was written and where much of the associated teaching occurred. For the most part, that is the land of the Turrbal and the Yuggera people and we pay respect to their Elders, past, present, and emerging. We recognise that passing on knowledge and building new understanding has occurred here for many thousands of years, and that we are just a small, recent part of a much bigger story in this place.

We commend the people of Papua New Guinea who welcomed us to their country. Every place on our planet has evolved over time and, for thousands of years, the custodians of all lands have successfully integrated and applied their knowledge of what we now refer to as science, technology, engineering, and mathematics. Some of this knowledge has passed on from one generation to the next. We are thankful to the PNG teachers who have shared some of this knowledge with us. We have learnt a great deal along the way—thank you very much.

We are indebted to the Australian Government for giving QUT the opportunity to deliver this project. The support provided by the Department of Foreign Affairs and Trade in Papua New Guinea has made an enormous difference to the success of this project. Extraordinary steps were taken to ensure that participants had the time and access needed to be successful when circumstances changed. We are also thankful to the National Department of Education in Papua New Guinea for their high level of commitment throughout the project. Thank you also to Dr Uke Kombra (Secretary for Education) and Mr Benny Rayappan (National Director of Secondary Education) for their ongoing support. Further, the assistance of many school administrators and teachers' families was critical to the success of this project.

We also acknowledge the United Nations for their development of their inspirational Sustainable Development Goals (https://www.un.org/sustainabledevelopment/sustainable-development-goals/) which have guided the design of our teaching and the implementation of STEM into the schools of our participating teachers. We are grateful for their permission to include the SDG icons in this book while acknowledging that the content of this publication has not been approved by the United Nations and does not reflect the views of the United Nations or its officials or Member States.

While a significant part of this project was riddled with challenges due to the COVID-19 pandemic, we were able to navigate our way through the complexities because of the high-quality support that was provided by the

Australia Awards team in Port Moresby. For this assistance, we are particularly thankful to Rebecca Gitsham (Facility Director), Simon Kaldy (Capacity Development Manager Higher Education and Short Course Awards), John Steer (Adviser, Short Course Awards), Baro Lawrence (Assistant Manager), Raymond Harold and Clara Numbasa (Senior Co-ordinators) for their unwavering support. We also acknowledge the support of David Kozar, Nelson Salangsang, and the International Projects Team at QUT for managing the project.

We have been extremely fortunate to work with some exceptionally talented and hard-working teachers across the three cohorts of the Graduate Certificate in STEM Education. We were delighted to hear how they have been applying their new knowledge and skills within their contexts. This has created new understandings for us in terms of how integrated STEM can be applied in diverse learning environments. The 18 narratives shared by 16 teachers in this book have taken our understandings to a new level and have played a role in defining STEM Education well beyond its component disciplines. We thank our teacher-authors for the time and effort put into documenting their experiences. We are grateful to the schools and students who enabled them to implement their ideas.

In conclusion, we extend our thanks to the QUT team for coming along on this journey. Their insightful contributions, as part of the teaching and writing teams, made an enormous difference to this project. We are grateful for their time, effort, and camaraderie which has kept this book project rolling.

Vinesh Chandra
Margaret Lloyd
Dann Mallet

Contents

Editors and Contributors

About the Editors

Vinesh Chandra is an Associate Professor and the STEM Subject Area Co-ordinator at Queensland University of Technology (QUT). In 2024, he was awarded an Australian Award for University Teaching (AAUT) for his contributions to advancing STEM education in Australia and overseas. He also served as one of the course leaders for the *PNG STEM Education Teacher In-service Project*. As well as being co-editor of this book, he has written Chap. 2 and contributed to Chaps. 3 and 5. Further to this, he, along with Dann Mallet, wrote the Introduction to and Review of Chap. 8 *Project-Based Learning*.

Margaret Lloyd is an Adjunct Professor, QUT, a Senior Fellow of the Higher Education Academy (HEA) and an Associate Fellow (Indigenous). She has won a National Teaching Excellence Award and is a National Teaching Fellow. She has had a long career in educational research and has worked extensively in academic professional development and international in-service teacher programs, particularly in S.E. Asia. Her editorial contribution to this book was mainly concerned with the design, collation and development of the teacher narratives presented and analysed in Part II: *Praxis—Teacher Narratives* and in grounding the teacher narratives within the Sustainable Development Goals.

Dann Mallet is Professor of STEM Education and Professor of Mathematics at Queensland University of Technology. He is experienced in academic and curriculum team leadership and mentoring, as well as extensive experience as a university Mathematics and STEM educator at all levels. Dann has initiated and led significant international education aid projects in the Indo-Pacific region, and has been recognised through numerous awards and nominations, including, most recently, a Gold in the QS Wharton Awards and High Commendation in the IEAA awards. Aside from his role as co-editor, he has contributed to Chaps. 3 and 4. Along with Vinesh Chandra, he wrote the Introduction to and Review of Chap. 8 Project-Based Learning.

Contributors

Gideon Aku Utmei Secondary School, Kokopo, Papua New Guinea

Terri Bourke School of Education, Queensland University of Technology, Brisbane, Queensland, Australia;
Kilakila Secondary School, Port Moresby, Papua New Guinea

Christopher N. Blundell School of Education, Queensland University of Technology, Kelvin Grove, QLD, Australia

Vinesh Chandra Faculty of Creative Industries, Education, and Social Justice, School of Education, Queensland University of Technology, Kelvin Grove, QLD, Australia

Charisse Farr School of Education, Queensland University of Technology, Kelvin Grove, QLD, Australia

Manasseh Kui Fena Wawin National School of Excellence, Lae, Papua New Guinea

Danielle Gordon School of Education, Queensland University of Technology, Brisbane, Queensland, Australia

Senka Henderson School of Education, Queensland University of Technology, Kelvin Grove, QLD, Australia

Douglas Hopkos Bishop Leo Arkfeld Secondary School, Wewak, Papua New Guinea

Margaret Lloyd Faculty of Creative Industries, Education, and Social Justice, School of Education, Queensland University of Technology, Brisbane, Kelvin Grove, QLD, Australia

Dann Mallet Faculty of Creative Industries, Education, and Social Justice, School of Education, Queensland University of Technology, Kelvin Grove, QLD, Australia

Abel P. Mambo Aiyura National School, Kainantu, Papua New Guinea

Levesa Morumo Wapenamanda Four Square Secondary School, Wapenamanda, Papua New Guinea

Koni Wamzil N'Drihin National School of Excellence, Lae, Papua New Guinea

Anastasia Paraka Park Secondary School, Mount Hagen, Papua New Guinea

Paula Niandros Polume Queensland University of Technology (QUT), Brisbane, Australia;
ECOM Secondary School, Lorengau, Papua New Guinea

Matthew Readette Queensland University of Technology, Kelvin Grove, QLD, Australia

Anna Semese Markham Valley Secondary School, Morobe Province, Papua New Guinea

Pauline Suvi Sogeri National School of Excellence, Sogeri, Papua New Guinea

Lonah Tombele Kilakila Secondary School, Port Moresby, Papua New Guinea

Susan Totori Kabiufa Adventist Secondary School, Goroka, Papua New Guinea

Raphael Ulka Gumine Secondary School, Kudiawa, Papua New Guinea

Jesse Waison Kerevat National High School, Kerevat, Papua New Guinea

Elias Wranga Cross Country School of Excellence, Kiunga, Papua New Guinea

Andy Yeh School of Education, Queensland University of Technology, Kelvin Grove, QLD, Australia

Debbie Zaccias Bumayong Lutheran Secondary School, Lae, Papua New Guinea

Abbreviations

AAPNG	Australia Awards Papua New Guinea
CK	Content Knowledge
CRI	Cocoa Research Institute
DFAT	Department of Foreign Affairs and Trade (Australian Government)
ENBP	East New Britain Province
ERT	Emergency Remote Teaching
LLG	Local Level Government
MTDP	Medium Term Development Plan
NARI	National Agriculture Research Institute
NGI	New Guinea Islands
NSoE	National School of Excellence
ODA	Official Development Assistance
PASS	PNG-AUS Secondary School Partnership
PCK	Pedagogical Content Knowledge
PGK or K	Kina
PK	Pedagogical Knowledge
PLC	Professional Learning Community
PNG	Papua New Guinea
PNG-AUS	Papua New Guinea-Australia
PNGDoE	Papua New Guinea Department of Education
PNGMTDS	PNG Medium Term Development
QUT	Queensland University of Technology
SBC	Standard-Based Curriculum
SBE	Standards-Based Education
SDG	Sustainable Development Goals
STEAM	Science, Technology, Engineering, Arts and Mathematics
STEM	Science, Technology, Engineering and Mathematics
TK	Technological Knowledge
UN	United Nations

List of Figures

List of Tables

Introduction to the Book

1

Vinesh Chandra, Margaret Lloyd, and Dann Mallet

Abstract

Simon Sinek's Golden Circle is a leadership and marketing model that can be applied in various contexts. At the core of his model is the guiding question of why, followed by how, and then what. Using this framework, this chapter explores three key questions: *Why was this book written? How is the book organised?* and *What does the book aim to achieve?* To provide context, the chapter also offers a brief background on Papua New Guinea, its education system, and the importance of teachers' professional development. Australia's role in enhancing the educational opportunities of PNG citizens is also discussed, with reference to the country's *Schools of Excellence Policy*, enacted in 2020, which emphasises STEM education as a key priority.

Keywords

Integrated STEM education · Papua New Guinea · Teacher in-service

Leadership expert Simon Sinek's TED talk entitled "How great leaders inspire action" has been viewed more than 67 million times (TEDx Talks 2019). In his presentation, Sinek shared his Golden Circle model and outlined why the "why", the "how", and the "what" questions are essential and why they need to be dealt with in this order when explaining a novel product or idea. We have used Sinek's model to explain the "why", the "how", and the "what" of this book. We have also provided a background to the "PNG STEM Education Teacher In-service Project" that was a key element of the context of this book.

1.1 Why Was This Book Written?

For decades, stakeholders with a vested interest in the education of future generations have talked about STEM Education. Proponents often framed the conversations around the need for citizens to be competent in Science, Technology, Engineering, and Mathematics. A workforce qualified in STEM knowledge and skills could potentially add significant value to a country's economy

V. Chandra (✉) · M. Lloyd · D. Mallet
Faculty of Creative Industries, Education, and Social Justice, School of Education, Queensland University of Technology, Kelvin Grove, QLD, Australia
e-mail: v.chandra@qut.edu.au

M. Lloyd
e-mail: mm.lloyd@qut.edu.au

D. Mallet
e-mail: dg.mallet@qut.edu.au

V. Chandra et al. (eds.), *Integrated STEM Education in Papua New Guinea*, Sustainable Development Goals Series, https://doi.org/10.1007/978-981-95-0828-0_1

and prosperity. Good job opportunities existed for those with STEM knowledge and skills. However, over the years, the uptake of these subjects in schools and universities, has declined. This is often attributed to how STEM subjects are taught. While many concepts in STEM are connected, the subjects are taught in silos. Consequently, students do not understand why they are learning what they are learning. One way to address this is by integrating STEM. Through such an approach, concepts in these subjects can be applied to tackle real-world projects, problems, and inquiries. In addition, students can use 21st-century skills, viz., critical and creative thinking, collaboration, and communication skills.

While integrating STEM ideas has merit, the real challenge for schools and teachers is how such an approach can be implemented. The literature (e.g. Shernoff et al. 2017) identifies the challenges that must be tackled before an integrated approach to teaching STEM can be fully realised. These include the training of teachers and examples of how such an approach can be delivered. This book was written to highlight how school systems can tackle these challenges. It presents a case study of how a group of teachers from Papua New Guinea were upskilled in integrated STEM education through the Graduate Certificate in STEM Education delivered by the Queensland University of Technology (QUT), and, in particular, the academic authors of this book. It also provides examples of how teachers apply their new knowledge and skills in their contexts by tackling real-world challenges and needs that are aligned with the Sustainable Development Goals (United Nations [UN], n.d.). For academics, such an approach provides them with evidence of how what they are teaching are being implemented by teachers. Through these connections, academia can sharpen its strategies for delivering teacher in-service courses on integrated STEM Education. A significant flaw within Overseas Development Aid (ODA) funded short courses is that it is assumed that teacher professional development will improve teacher quality and student learning. In a rigorous 5-year mixed-methods study conducted in Timor Leste, Vanuatu and Lao PDR, it was found that there are no "robust monitoring and evaluation systems in place that systematically capture outcome-level data to provide evidence of quality and impact" (Hollingsworth et al. 2023, p. 10). The project reported in this book, supported by the Australian Government, has demonstrated how monitoring and evaluation can be enacted to ascertain the impact of professional learning initiatives on teacher practice.

1.2 How Is the Book Structured?

This chapter provides an overall introduction to the book which is divided into two main parts. Part I, encompassing Chaps. 2– 5, will focus on professional learning and leadership associated with the PNG STEM Education Teacher In-service Project. Part II, encompassing Chaps. 6–9, offers 18 teacher narratives by 16 individual teacher authors organised into four categories (Fig. 1.1).

Parts I and II are collectively representative of stages of the Experiential Learning Cycle (Kolb 1984) which has guided our practice (see Chap. 2). And finally, in conclusion, Chap. 10 will draw together the elements of the book and provide guidance for future practice.

Part I of this book has four chapters. These cover our Philosophy and Background to Course Design (Chap. 2), the design and delivery of the course and their impact on teachers drawn from targeted research (Chaps. 3 and 4), and the global implications for professional learning and leadership in STEM (Chap. 5).

Part II presents selected narratives of the teachers who undertook this postgraduate study with reviews of the narratives provided by university academics who were involved with the course delivery. Teachers, as authors, initiated projects which addressed pressing demands in their schools. They investigated and wrote about what was most important to them and their communities. The STEM-related issues may have been the professional learning of their colleagues or the real problems being tackled in their daily lives. What they planned was dependent on academic

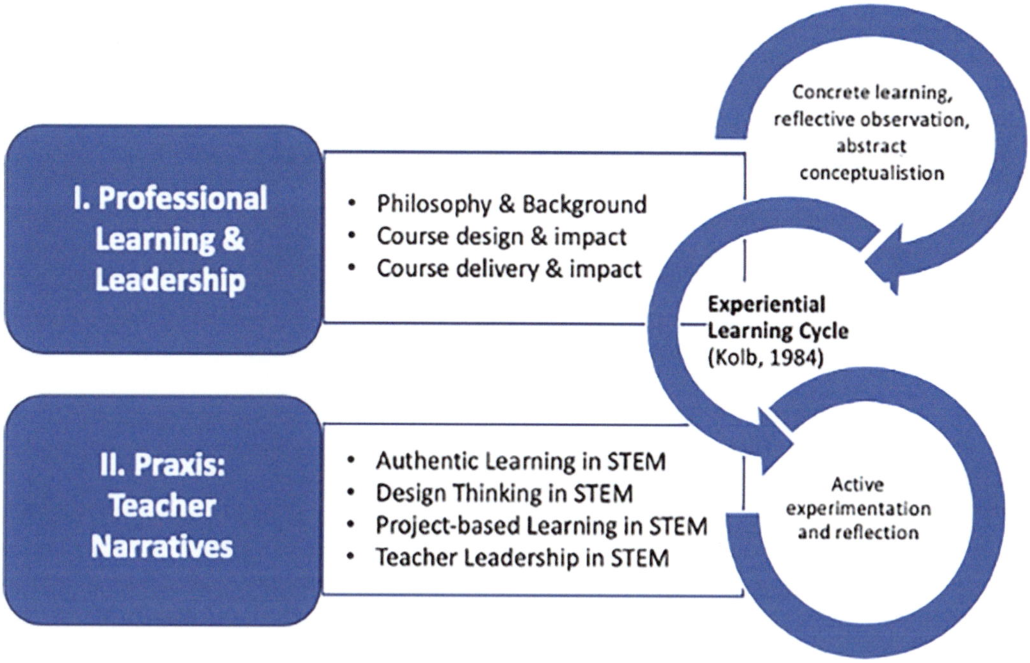

Fig. 1.1 Structure of book

and environmental priorities and circumstances. Their narratives are grouped into four discrete contexts of STEM education: Chap. 6 focusses on Authentic Learning, Chap. 7 is concerned with Design Thinking, Chap. 8 presents Problem-Based Learning, while Chap. 9 delves into Teacher Leadership.

Chapter 10 provides a conclusion to the book by describing what the teachers have achieved in terms of integrated STEM Education in their contexts while also targeting the Sustainable Development Goals (UN, n.d.). It also highlights how the "PNG STEM Education Teacher In-service Project" added new knowledge to the literature. In sum, this final chapter offers a way forward for STEM teacher learning and practice in developing countries and elsewhere. Appendices have been provided to show how the narratives relate to the Sustainable Development Goals (Appendix A) and to broader curriculum objectives which have been achieved through STEM activities (Appendix B).

1.3 What Does the Book Hope to Achieve?

This book is unique for several reasons. Firstly, it documents what in-service teachers have learnt in a professional learning course and how they have applied their new STEM Education knowledge and skills in their contexts. The latter is presented through eighteen teacher narratives. Each narrative connects with Sustainable Development Goals and is related to a discrete STEM Education context.

Secondly, the book presents the findings of an investigation which explored the impact on the teachers of the in-service postgraduate course design and delivery. Additionally, this book is novel because it was developed with the input of the university academics who delivered the course as well as the practising teachers who were enrolled in the program. Fourthly, the book documents the impact of a scholarship program funded by the Government of Australia.

Finally, the book demonstrates Kolb's Experiential Learning Cycle (Kolb 1984). The cycle offers a four-step learning process: concrete learning, reflective observation, abstract conceptualisation, and active experimentation. Part I of this book is about teachers learning and engaging through Kolb's first three steps (concrete learning, reflective observation, and abstract conceptualisation). Part II presents teacher narratives of active experimentation and reflection, the last step in Kolb's learning process (see Fig. 1.1).

1.4 Background to PNG STEM Education Teacher In-Service Project

Papua New Guinea (PNG) is a developing country located in the southwestern Pacific Ocean. It occupies the eastern half of the island of New Guinea, the Bismarck Archipelago, Bougainville, Buka, and some six hundred smaller islands and atolls. Papua New Guinea is known for its diverse geography including tropical rainforests, mountains, coral reefs, and active volcanoes. It is home to a diverse range of flora and fauna, some of which are unique to the region. The capital and largest city of Papua New Guinea is Port Moresby, located on the southern coast of the country. PNG has a population of about 10.6 million and about 85% of the population live in rural areas. It is considered one of the most linguistically diverse places in the world with over 850 languages spoken, many of them with fewer than 1000 speakers. The official languages are English, Tok Pisin (a Creole language), and Hiri Motu. PNG has a rich cultural heritage, with traditional customs and rituals still practised in many communities. The country is known for its unique tribal art, including intricate carvings, masks, weavings, and sculptures.

Schooling in Papua New Guinea has followed a 3-6-4 structure—three years of elementary school, six years of primary education, and four years of secondary education (Nunn and Nelson 2019). This structure includes merit-based exams to transition from primary to the first two years of secondary school followed by another exam to qualify for Grades 11 and 12. PNG is now transitioning to a model of 1-6-6, 1 year of early child education, six years of primary education, and 6 years of secondary education (Goro 2021) with progression examinations abolished.

The delivery of programmes in schools is guided by two curriculum documents. The first of these is the Standards-Based Curriculum (SBC) for schooling from Prep to Grade 12 (Oko 2021) which came into effect in 2014. The second is the STEM curriculum introduced in 2020 which is only available for Grades 11 and 12 students enrolled in the six Schools of Excellence located at Kerevat, Aiyura, Sogeri, Passam, Port Moresby, and Wawin.

Primary education is compulsory in PNG, but it is not accessible for all children. While efforts have been made to improve access to education, significant challenges remain. Many children, especially in rural and remote areas, do not have access due to the lack of infrastructure and long distances to travel. As a result, the enrolment rate is low, particularly at the secondary level.

Disparities such as these impact the overall quality of education across the country. English is the language of instruction in PNG schools, but many students have different mother tongues. This language barrier can hinder effective learning and communication in the classroom. Further, gender disparity in education is also a significant issue. Girls, especially in rural areas, face barriers such as early marriage, cultural norms, and limited access to educational opportunities. Efforts made to promote gender equality have been met with mixed success.

The Government of Papua New Guinea has recognised the importance of education and has implemented various initiatives to improve its education system. For example, the government implemented the Tuition Fee Free (TFF) Policy in 2012 to provide fee-free education for primary and lower secondary levels. However, due to variations in implementation and management practices, parents have had to co-contribute towards their children's school expenses. Teacher quality is also seen as a barrier to advancement and the government's efforts to reform the education sector have been hampered by a lack of a "clear roadmap and funding for the upskilling of teachers" (Nunn and Nelson 2019, p. 2).

The below-par performance of many students in Mathematics at the secondary school level motivated Oko (2021) to examine the factors—school, teacher, student—and their interrelationships on Grades 10 and 12 students' results in the subject. Oko's study involved 729 students and 41 teachers and data was gathered through a quantitative-dominant mixed-method approach. His study proposed that teachers needed refresher courses on mathematics content and pedagogy so that they develop "confidence … and enrich themselves with the necessary skills to assist students to learn mathematics effectively" (p. 340). His study also found that student-centred and mixed teacher- and student-centred approaches made a significant difference in learning outcomes, compared to exclusively teacher-centred teaching methods. He believed that student-centred pedagogies promoted:

> … effective teaching environment(s) and create(d) positive relationship(s) between the students and the teachers … [and that] encouraging dialogue approaches delivered deeper and more meaningful learning and also developed respect that recognise(d) individual students' value, importance and contribution. (p. 340)

Based on this finding, Oko (2021) concluded that "more professional development related to these teaching pedagogies should be encouraged in all schools to upgrade teachers' professional capabilities" (p. 341).

For decades, Australia has played a significant role in advancing and supporting PNG's education agenda including teachers' capabilities through numerous initiatives. For example, since the early 1950s, Australia has provided more than 12,000 scholarships to Papua New Guineans for tertiary study in Australia and in Papua New Guinea (PNG), as well as academic or professional short courses. From 2016 to 2023, the Australia Awards Papua New Guinea (AAPNG) Program continued the tradition by supporting projects to an estimated value of $170 million (Australian Government 2021). In 2020, PNG became a world leader by launching the "Schools of Excellence Policy", with a strong national focus on Science, Technology, Engineering, and Mathematics (STEM). The rationale was to "produce students who are knowledgeable, highly educated, and agile in their abilities to comprehend scientific concepts and improvise to create innovative tools to solve everyday problems" (PNG PM Marape 2020, para.12).

Papua New Guinea has been ambitious in embracing transdisciplinary STEM education nationally taking a giant leap towards improving opportunities for its citizens. However, the country did not have teachers trained in STEM Education to deliver this ambitious plan, and it lacked the human and financial capital to effect the transformation. Given this position, the PNG Government approached the Australian Government seeking to introduce STEM Education by training high school teachers from the country under the Australia Awards Short Courses programme. As a result, the "PNG STEM Education Teacher In-service Project" was born to upskill numerous groups of PNG school teachers. This book is dedicated to the teachers who participated in the programme, and it documents aspects of their learning and how they are applying their new knowledge and skills in their contexts.

We are guided by the belief, attributed to William Butler Yeats, that "education is not the filling of a pail, but the lighting of a fire". We hope that our collective efforts in editing and writing this book spark new thoughts and ideas. We further hope that the experiences of the contributors to this book will light the fire of integrated STEM Education to inspire teaching and learning, wherever in the world you are and whoever you teach.

References

Australian Government (2021) Australia awards Papua New Guinea (AAPNG) Review Report. https://www.dfat.gov.au/publications/development/australia-awards-papua-new-guinea-review-report

Goro J (2021) Is it the right time to implement the new 1+6+6 basic education system in Papua New Guinea? Spotlight 14(4). The National Research Institute Papua New Guinea

Hollingsworth H, Wong D, Cassity E, Goundar P (2023) Teacher development multi-year study series: report

highlights. Multi-country report. Aust Counc Educ Res. https://research.acer.edu.au/eas/72
Kolb DA (1984) Experiential learning: Experience as the source of learning and development (Vol. 1). Prentice-Hall
Nunn RM, Nelson G (2019) PNG: Has education policy reform worked. https://www.lowyinstitute.org/the-interpreter/png-has-education-policy-reform-worked
Oko J (2021) An investigation into factors affecting Grade 10 and 12 students' mathematics performance in Port Moresby, Papua New Guinea (PNG) (Doctoral dissertation). https://hdl.handle.net/2440/134190
PNG PM Marape launches Schools of Excellence Policy (2020, June 15) Papua New Guinea Today. https://news.pngfacts.com/2020/06/png-pm-marape-launches-schools-of.html
Shernoff DJ, Sinha S, Bressler DM, Ginsburg L (2017) Assessing teacher education and professional development needs for the implementation of integrated approaches to STEM education. Int J STEM Educ 4:1–16. https://doi.org/10.1186/s40594-017-0068-1
TEDx Talks (2019, November). How great leaders inspire action | Simon Sinek | TedxPugetSound. Ted Talk. https://www.ted.com/talks/simon_sinek_how_great_leaders_inspire_action/c
Thibaut L, Ceuppens S, De Loof H, De Meester J, Goovaerts L, Struyf A, Boeve-de Pauw J, DehaeneW, Deprez J, De Cock M, Hellinckx L, Knipprath H, Langie G,Struyven K, Van de Velde D, Van Petegem P, Depaepe F (2018) Integrated STEM education: a systematic review of instructional practices in secondary education. Eur J STEM Educ 3(1):02. https://doi.org/10.20897/ejsteme/85525
United Nations (UN) (n.d.) Sustainable development goals. https://www.un.org/sustainabledevelopment/sustainable-development-goals

Vinesh Chandra is an Associate Professor and the STEM Subject Area Co-ordinator at Queensland University of Technology (QUT). In 2024, he was awarded an Australian Award for University Teaching (AAUT) for his contributions to advancing STEM education in Australia and overseas. He also served as one of the leaders for the *PNG STEM Education Teacher In-service Project*. For this project, the team was awarded a Gold in the QS Wharton Awards and a High Commendation in the IEAA awards. In addition to being co-editor of this book, Associate Professor Chandra authored this chapter and contributed to others. He has also co-authored other books on STEM education, including the award-winning "STEM Education in the Primary School: A Teacher's Toolkit" published by Cambridge, and "My STEM Workbook" for primary school students.

Margaret Lloyd is an Adjunct Professor, QUT, a Senior Fellow of the Higher Education Academy (HEA) and an Associate Fellow (Indigenous). She has won a national Teaching Excellence Award and is a National Teaching Fellow. She has had a long career in educational research and has worked extensively in academic professional development and international in-service teacher programs, particularly in S.E. Asia. Her editorial contribution to this book was mainly concerned with the design, collation and development of the teacher narratives presented and analysed in Part II: *Praxis—Teacher Narratives* and in grounding the teacher narratives within the Sustainable Development Goals.

Dann Mallet is Professor of STEM Education and Professor of Mathematics at Queensland University of Technology. He is experienced in academic and curriculum team leadership and mentoring, as well as extensive experience as a university Mathematics and STEM educator at all levels. Dann has initiated and led significant international education aid projects in the Indo-Pacific region, and has been recognised through numerous awards and nominations, including, most recently, a Gold in the QS Wharton Awards and High Commendation in the IEAA awards. Aside from his role as co-editor, he has contributed to Chaps. 3 and 4. Along with Vinesh Chandra, he wrote the Introduction to and Review of Chap.8 *Project-Based Learning.*

Part I
Professional Learning and Leadership

Part I of this book, through its four chapters, has a threefold purpose. First, it describes the philosophy of STEM Education and the current thinking into how in-service teachers can gain the experience and mindset needed to become effective teachers of STEM. Second, through case study research, it investigates how our *design* of the postgraduate STEM Education course impacted on the teachers' understanding and knowledge of integrated STEM Education. Third, it is about understanding how our approach to the *delivery* of the course and external factors affected teachers' learning outcomes.

The teachers who undertook the Graduate Certificate in STEM Education, all from Papua New Guinea (PNG), are here considered as non-traditional learners because most were not only experiencing multimodal learning for the first time, they were also studying from home or had completed parts of their course in an intensive mode by temporarily relocating to capital cities.

Between 2020 and 2023, the STEM Education course was delivered to three cohorts of teachers from PNG. Traditionally, Australia Awards Short Courses are designed to be delivered on campus at an Australian university with cohorts travelling to Australia and living near the university for the duration of their study (Australian Government, n.d.). This was the original plan for the three cohorts of teachers undertaking the STEM Education course. The original plan was to deliver the course in two six-week face-to-face blocks, separated by a few months, with teachers completing half of the postgraduate certificate course in each block. In March 2020, Cohort 1 (26 teachers) commenced their studies at our campus in Australia. Two weeks into their study, the teachers were directed to return home to PNG, due to imminent border closures related to the COVID-19 pandemic.

Rapidly changing and uncertain government mandates due to the pandemic played havoc on our course delivery strategies. The proverb "necessity is the mother of invention" drove our team to explore innovative ways in which the course could be delivered online to teachers, some in very remote locations in PNG. This presented us with an excellent opportunity to reimagine how we would plan and develop learning intentions through the online delivery model.

In September 2020, Cohort 2 comprising a further 26 PNG teachers enrolled in the course. Because of ongoing disruptions caused by COVID-19

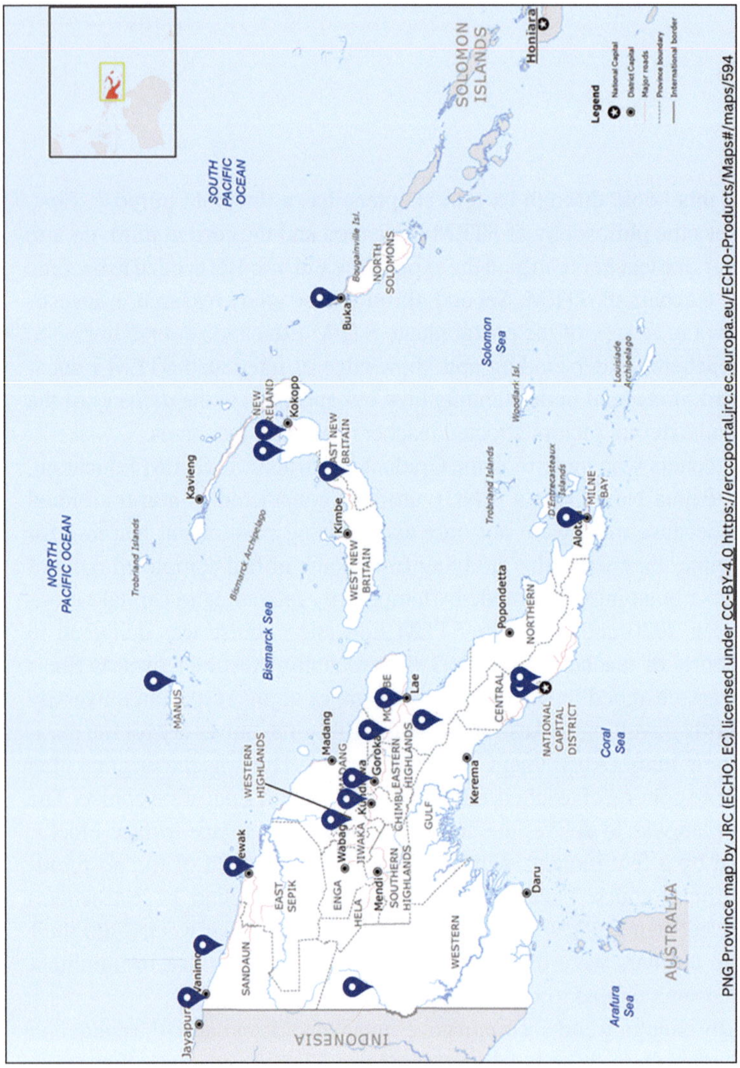

Fig. 1 Provincial map of Papua New Guinea indicating the distribution of teachers enrolled in the Graduate Certificate in STEM Education. Image source:https://erccportal.jrc.ec.europa.eu/ECHO-Products/Maps#/maps/594CC-BY-4.0

during 2020-21, the course could only be delivered in an online mode to both cohorts. A third cohort of 25 more PNG teachers (Cohort 3) enrolled in the course in June 2022. For this group of teachers, half of the course was delivered in Port Moresby and the other half at our campus in Brisbane. In all, a total of 77 teachers were enrolled in the STEM Education course.

The teachers had been selected from various regions across PNG ensuring broad representation and dissemination. As shown on the map in Figure 2, some of the teachers were based in schools close to cities like Port Moresby and Lae while some were from smaller townships such as Kiunga, Vanimo and Goroka. Other teachers hailed from islands like Manus, Buka, and New Britain which have very small-town centres. From the capital Port Moresby, most of these places could only be reached by air or sea. Thus, teachers' school locations varied from urban to rural and very remote.

As is the case in many parts of the developing world, the location of schools and teachers' homes often determines what teachers can and cannot do in the classroom. For example, internet connectivity and electricity supply are severely hampered in geographically isolated locations with smaller populations such as Alotau when compared to the capital, Port Moresby. Such issues can have a significant impact on how teachers engage in professional learning activities, especially when they are delivered online.

A total of 75 teachers completed the course (from 77 initial participants). All teachers who were enrolled in the course had extensive classroom experience ranging from 5–10 years (26%), 11–15 years (46%), and more than 15 years (28%). Their teaching qualifications ranged from a diploma (15%) to bachelor's degree (85%). Most of the teachers (85%) were from schools where their colleagues were all qualified to teach in their respective STEM subjects. About 60% indicated that their class sizes could comprise up to 50 students. However, 18% of the teachers also indicated that class sizes could exceed 50 students. While 85% of the teachers were from schools with 1 to 5 science laboratories, 15% were from schools with no laboratories at all. Very few schools had specialist teaching areas such as home economics, manual arts, or music rooms. However, they could all access outdoor areas for teaching purposes such as a farm, ponds, and rivers adjacent to the schools.

Access to the internet in the schools varied. It was also interesting to note that different schools had different policies on who could access the internet, including (a) sole use by the school's administration, (b) teachers given data use allowance, (c) students given a data use allowance, (d) teachers and/or students being asked to pre-pay for data use, (e) teachers allowed to use the internet for lesson preparation only, and (f) teachers and students allowed to use data for classroom work. For most teachers, the digital technologies in the school were over 5 years old. Desktop computers were the most popular technology and about 50% of schools had data projectors. Only a handful of schools had newer technologies such as 3D printers, scanners, and electronic libraries. Electricity supply to schools varied as follows: no problems with electricity (20%), occasional outage (at least once a month) (41%), frequent outages (at least once a week) (23%), and regular outages (at least once a day) (12%). The teacher narratives, presented in Part II, further highlight restrictions to other basic utilities such as water and equipment.

A pathway for enhancing the quality of education is through creating opportunities for the professional learning of teachers and encouraging them to become leaders. This opportunity should be available to all teachers irrespective of where they are located. Our work in PNG delivered some outstanding outcomes (reported and analysed in Part II). We begin Part I by reviewing the literature relating to teacher learning in STEM and in articulating our own philosophy to guide our course design (Chapter 2).

Further to this, to gain insight into what the PNG teachers thought about the course design we developed and the course delivery that was imposed upon us, we set up a research project and invited the teachers to participate. Two chapters in this section target specific research questions that informed our investigation.

Chapter 3—*Course design and its impact on teachers*

- How did the course design impact on teachers' knowledge on integrated STEM Education?

Chapter 4—*Course delivery and its impact on teachers*

- How did the modes of course delivery impact on teachers' learning?

Part I will conclude with a discussion of our findings and observations through the conduct of our course (Chapter 5). We believe that, although conducted in a specific location for a specific group of teachers, our observations and research findings provide broader implications for practice of STEM Education in other parts of the world.

section

ReferenceAustralian Government. (n.d.). Australia Awards. https://www.dfat.gov.au/people-to-people/australia-awards

2 Philosophy and Background to Course Design

Vinesh Chandra

Abstract

This chapter outlines the philosophy behind the design of the Graduate Certificate in STEM Education course that the PNG teachers were enrolled in. The course consisted of eight subjects, divided into two phases: the first focused on developing pedagogical content knowledge (PCK), and the latter on applying that knowledge in realworld contexts. We adopted ideas from Jerome Bruner (spiral curriculum) and David Kolb (experiential learning theory) in course design. We used them in the "Cyclical Experiential Learning model", which was underpinned by four phases: concrete experience, reflective observation, abstract conceptualisation, and active experimentation.This model effectively facilitated the progressive development of the teachers' knowledge, understanding, and skills while achieving the key learning intentions across the eight subjects.

Keywords

PCK · Spiral curriculum · Experiential learning theory · Cyclical Experiential Learning model · Graduate Certificate in STEM Education

With growing interest in integrated STEM Education internationally (Jorgensen and Larkin 2018; Li et al. 2019), there is a need for a paradigm shift within school systems regarding how the core business of teaching and learning occurs in these subjects. The lion's share of implementing an integrated approach falls on in-service teachers. However, many feel "ill-equipped to implement integrated STEM" (Shernoff et al. 2017, p. 13). Recognising and creating changes in teachers' mindsets and practices are therefore vital. For these changes to occur, there is a need for "educational opportunities that are more substantial than isolated in-service workshops (p. 13).

These researchers have also suggested a need for training that enables teachers to work outside their "specialisation so that they can integrate multiple learning goals from different content areas" (p. 13). Effective professional learning programmes also need to be developed so that teachers embrace ideas outside their specialisations , such as engineering design and project-

V. Chandra (✉)
Faculty of Creative Industries, Education, and Social Justice, School of Education, Queensland University of Technology, Kelvin Grove, QLD, Australia
e-mail: v.chandra@qut.edu.au

V. Chandra et al. (eds.), *Integrated STEM Education in Papua New Guinea*, Sustainable Development Goals Series, https://doi.org/10.1007/978-981-95-0828-0_2

based learning that are strongly underpinned by constructivist, inquiry-based, and project-based pedagogy.

The idea of implementing integrated disciplinary approaches to teaching has been around for a few decades. For example, in the 1990s, the National Science Foundation mooted the idea of integrating science, mathematics, and technology (Salinger and Zuga 2009). In the same vein, the Committee on K-12 Engineering Education highlighted the use of "engineering concepts and skills to leverage the natural connections between STEM subjects" (Katehi et al. 2009, p. 12). The committee believed that the siloed teaching approach of STEM subjects had "impeded efforts to increase student interest and improve performance in science and mathematics" (p. 12).

For teachers, an integrated approach to teaching STEM education entails bringing together knowledge and skills from the four disciplines and presenting them to students through hands-on learning activities. The activities relate to such pedagogical approaches as project or problem-based learning, inquiry-based learning, authentic learning, or design thinking. Teachers can plan activities by drawing upon the concepts from two, three or four disciplines, and the knowledge and skills from the identified disciplines do not need to be in equal or predefined proportions. The majority of teacher narratives presented in Part II have been loosely grouped by pedagogy. Here, we have descriptions of activities in *Authentic Learning* (Chap. 6), *Design Thinking* (Chap. 7), and *Project-Based Learning* (Chap. 8). A fourth group relating to *Teacher Leadership* (Chap. 9) reports on teacher adoption and advocacy of these pedagogies.

Through integrated STEM activities, students are expected to develop critical and creative thinking, collaboration, and communication skills, all of which are desirable qualities of 21st-Century citizens (Guzey et al. 2020). Teachers must also understand that students do not always use their STEM skills in isolation. They have other "preconceptions and misconceptions" that can influence their learning (Shulman 1986, p. 9). For the current generation of school students, the internet has opened the door to new ways of learning and knowing. This technology can also create pathways for developing the students' preconceptions and misconceptions about our world. The obvious question that then follows, is how do we train teachers to deliver teaching and learning outcomes under some of these conditions. It is also important to note that most of the in-service teachers in the current workforce went through their training at a time when integrated STEM was an unknown quantity. In the case of secondary school teachers, particularly in the senior school, disciplinary knowledge and expertise is often confined to one discrete discipline (or, at best, two disciplines).

Several studies have identified the limited preparation and the lack of qualified teachers as one of the key barriers to the delivery of an integrated STEM approach in classrooms (e.g., Ejiwale 2013). In a study entitled *Assessing Teacher Education and Professional Development Needs for the Implementation of Integrated Approaches to STEM Education*, Shernoff et al. (2017) noted that there was a "universal belief that current teacher education is wholly inadequate and would need to be considerably rethought and revamped for integrated STEM to flourish in schools" (p. 13). Teachers needed pedagogical skills on how to bring together the content knowledge and skills across the four disciplines in their classrooms to robust activities that connected students' learning to the real world. Secondary school in-service teachers have a "highly unique need" because they have to step outside their traditional teaching specialities and embrace new pedagogical approaches (e.g., engineering design challenges), and how various parts can come together in an environment where they predominantly play the role of facilitator (Shernoff et al. 2017).

What teachers need are structured courses that develop their pedagogical content knowledge skills (Love and Hughes 2022). The design of the Graduate Certificate in STEM Education course which the PNG teachers were enrolled in addresses some of the issues raised in the literature (e.g., Gorozidis and Papaioannou 2014; Love and Hughes 2022; Shernoff et al. 2017; Thibaut et al. 2018).

In the Graduate Certificate in STEM Education offered at QUT, students complete eight subjects, each with a different focus relating to STEM Education. One of our priorities was to ensure that the course content was contextually relevant to teachers in the PNG context. According to the literature, a lack of relevance to local contexts together with a disconnect with "local cultural and educational circumstances" has historically impacted the quality of "pre-and in-service training" delivered by an "international donor community providing aid to less-developed countries" (Hardman et al. 2012, p. 833). To ensure contextual connectedness, all the subjects in the course were strongly focused on PNG while simultaneously maintaining academic rigour expected of a graduate-level degree.

To upskill teachers in this way, we believed that in-service training should develop their knowledge and skills enabling them to: (1) understand the rationale of integrated STEM; (2) identify and implement contextually appropriate projects, problems, and inquiries; (3) implement appropriate pedagogical and assessment approaches; and (4) design and implement effective classroom activities that target learning goals and practices from different STEM disciplines (Thibaut et al. 2018). These attributes framed the course design which was further informed through the cognitive development and learning theories of Jerome Bruner and David Kolb.

Bruner's (1960) theory postulated a spiral curriculum where the learner revisits a topic or theme throughout the course. The complexity of learning also increases over time. But more importantly, each "new learning" connects with the "old learning" (p. 39). Biesta (2019) reasoned that such an approach suited the design of teachers' professional learning and proposed that teacher education should:

> … start with the whole—that is to introduce the question as to what, in this situation, is an educationally desirable way forward—from day one, and keep expanding, transforming, deepening and just 'kicking around' this question throughout the whole course of study. In my view, this is not only a more honest way to let teachers and students encounter the 'essence' of the practice of education. It may also make it possible to constantly keep questions about the *how* and the *what for* 'in conversation,' thus deepening both students' understanding of the way in which *how*-questions impact on *what-for*-questions, and the way in which *what-for*-questions impact on *how*-questions. (p. 269)

David Kolb, the founder of the experiential learning theory, defined learning as "the process whereby knowledge is created through the transformation of experience. Knowledge results from the combination of grasping and transforming experience" (Kolb 1984, p. 41). Experiential learning relates to the intellectual growth of an individual which is linked with experience (Kolb and Kolb 2009). Thus, experiential learning combines experience, perception, cognition, and behaviour. By adopting such an approach in structured programmes, the following steps can deliver rich learning experiences for in-service teachers:

1. *Concrete Experience*: Teachers can engage in hands-on learning experiences relevant to their teaching context. They can observe other teachers in action, participate in classroom simulations, or try new teaching strategies in a safe and supportive environment.
2. *Reflective Observation*: After engaging in a concrete experience, teachers can reflect on what they learned through journaling, group discussions, or one-on-one coaching sessions. The goal is to help teachers identify what worked well, what didn't, and why.
3. *Abstract Conceptualisation*: Once teachers have reflected on their experiences, they can analyse and make sense of their observations by exploring theoretical frameworks or research-based strategies to inform their teaching practices. Finally, they can plan and develop ideas for implementing them in their contexts.
4. *Active Experimentation*: Teachers can experiment with new teaching strategies in their classrooms. For example, they can implement their ideas that draw upon what they learned in the previous stages of the learning cycle.

Based on the work of Jerome Bruner and David Kolb, we adopted a cyclical experiential learning

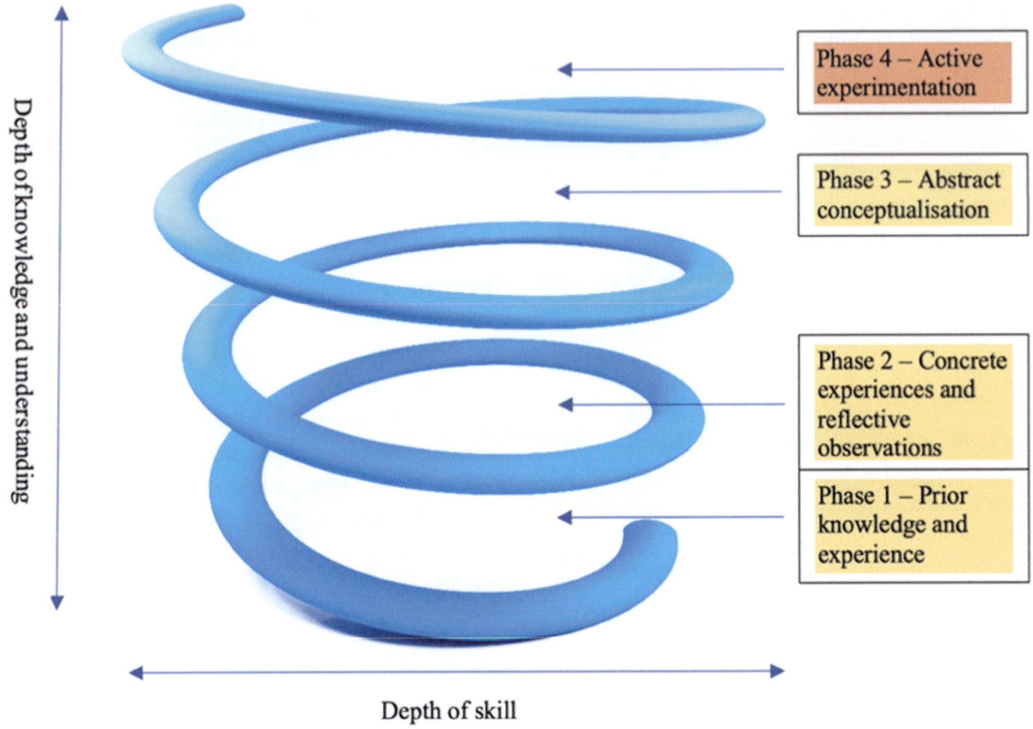

Fig. 2.1 Cyclical experiential learning model

model for in-service teacher professional learning in integrated STEM Education (Fig. 2.1).

Our cyclical experiential model similarly has four phases (Fig. 2.1). Phase 1 refers to the prior knowledge and experiences of the learners. The design of our course took into consideration the PNG teachers' teaching experiences and their professional qualifications and was influenced by their school and home contexts. We also noted that the PNG curriculum strongly emphasises traditional and cultural knowledge and practices. Within our team, we had STEM Education as well as international education experience, particularly in the Pacific Islands (e.g. Chandra 2019). The background and experience of the academic teaching team ensured a course design that accounted for the PNG context.

Teachers' full-time learning was interspersed with periods of full-time classroom teaching (aligning with the concrete experiences and reflective observations required of Phase 2). For each of the three cohorts of teachers, we were forced to adopt different models of delivery for the course learning experiences influenced by the prevailing conditions at the time (see Chap. 4 for complete details). For example, the course was delivered in three chunks or blocks of learning for one of the cohorts. Here, teachers would attend classes for a few weeks, complete some subjects of the course, return home to their teaching duties in schools, and then return for the next block of classes. This led to a three-way connection between concrete learning experiences in their STEM Education coursework, concrete experiences in teachers' own classrooms with their students, and reflective observations. The teachers were afforded a chance to not only gain concrete experiences in the course through hands-on activities and workshops, but they could also apply what they were learning in their own classrooms (see Fig. 3.1).

These experiences created numerous opportunities for reflection and deep thinking regarding what was realistically possible in their classrooms and contexts across PNG. These experiences and reflections fed into Phase 3 where teachers could engage in abstract conceptualisation and propose how they could apply their knowledge and skills. Teachers demonstrated their capabilities of these attributes through in-class activities and assessment tasks. Teachers were immersed in Phases 1, 2, and 3 of the cyclical experiential learning model in our course. Phase 4 allowed them to experiment with what they had abstractly conceptualised in Phase 3. They were able to test how their concepts played out in reality.

The vertical axis of the cyclical experiential learning model represents the depth of knowledge and understanding while the horizontal axis represents the depth of skill in applying integrated STEM practices in their classrooms (Fig. 2.1). Thus, as the teachers progressed through the phases, their knowledge and understanding of integrated STEM Education are enhanced. When they reach Phase 4, they can apply their knowledge and skills within their own individual contexts. However, as they become more knowledgeable, they also become more skilled as they continue to spiral upward. As skill grows so does the width of the spiral.

2.1 Course Content and Structure

The postgraduate course was designed with the practising teacher in mind. We accepted that course content, irrespective of modality, was perceived as the most critical factor in learning to "facilitate and reinforce the learning experience" (Castle and McGuire, 2010, p. 38). Our design and planning in the Graduate Certificate in STEM Education was meticulous. Each of the eight interconnected subjects targeted a different aspect of the critical explicit question revisited by the teachers through the course, viz., *How can I design and implement an integrated approach to teaching STEM in my context?*

Within each subject, the first three phases of the cyclical experiential learning model played a critical role. All had contextually focused learning intentions which collectively gave the teachers a range of opportunities to develop their knowledge and skills across STEM disciplines (Fink 2003). The first five subjects developed teachers' pedagogical content knowledge (PCK) in integrated STEM Education. The last three subjects focused on developing teachers' capacities to apply their PCK within their contexts (Table 2.1).

PCK was initially described as a "special amalgam of content and pedagogy that is uniquely the province of teachers" (Shulman 1987, p. 8). It is a specialised knowledge that teachers draw on while transforming subject matter or discipline knowledge into a more understandable form for students. Gess-Newsome et al. (2019) and Carlson et al. (2019) have respectively extended Schulman's idea to include "personal PCK" which develops this knowledge through teachers' failures and successes and is subject-specific. "Collective PCK" is the knowledge attributed to a group of teachers. Thus, a mathematics teacher may have their ways of teaching simultaneous equations, which might be quite different, but no less valid or effective, from how other school teachers might teach the concept.

The rationale for adopting PCK in subject design for the STEM Education course was that most teachers who enrolled in the course would have been trained in at least one STEM discipline. This was the case for all the PNG teachers enrolled in our course. They were mostly secondary teachers trained in one field (e.g., Mathematics, Physics, Chemistry, or Biology) (see Introduction to Part I). Hence, these teachers had a solid grounding in content knowledge (CK) in at least one STEM-related discipline. They also had teaching experience and therefore can be presumed to possess pedagogical knowledge (PK) meaning they could create teaching and learning environments for their students within their disciplines. In integrated STEM Education, PCK is where teachers need training because the focus is on more than one subject. PCK "cannot be simply imported from one subject matter to another" (Loughran et al. 2012, p. 7). It is also noteworthy that deep subject knowledge develops over time which can blur the boundary between content and pedagogy. Thus, a Physics teacher in a secondary school who has taught the subject for ten years might find it challenging to apply an

Table 2.1 Content of graduate certificate STEM education

Subject	Key learning intention	Assessment artefact	Course objective
1. STEM for 21st-Century learners	Understanding conceptual frameworks associated with integrated STEM in Education and reflection on personal experiences	Grant application	*Developing* pedagogical content knowledge in integrated STEM Education
2. Digital fluency for STEM learning	Exploring digital technologies as tools for learning that develop fluency in STEM teaching and learning	Digital portfolio	
3. Thinking in STEM	Understanding discipline-specific thinking and how it supports a connected approach to STEM teaching and learning	Digital portfolio	
4. STEM pedagogies	Developing knowledge of different pedagogies and how it can be implemented in STEM classrooms	Website	
5. Assessment and STEM learning	Developing an understanding of how assessment is enacted in integrated STEM activities	Classroom assessment package	*Applying* pedagogical content knowledge in integrated STEM Education to real-world contexts
6. Designing for connected learning	Developing and designing transdisciplinary curriculum and instruction aligned to target learning goals in STEM Education	Teaching plan	
7. STEM inquiry experience	Experiencing real-world applications and designing authentic learning that achieves the intentions of the STEM curricula	Digital portfolio	
8. Applied STEM in Education	Developing and presenting a strategic plan that has the potential to advance STEM Education in an education context	Action plan	

acquired "physics PCK" to a STEM classroom. We were mindful that this was exactly what we were expecting of the teachers enrolled in our course.

When designing courses for adults, Knowles' theory of andragogy is an important consideration (Weegar and Pacis 2012). Adults learn differently from younger students with the following being key considerations for adult learning: (a) adults need to know *why* they need to learn something; (b) adults need to learn experientially; (c) adults approach learning as problem-solving; and (d) adults learn best when the topic is of immediate value (p. 15). These aspects were fundamental to our STEM Education course design. Teaching and learning within the subjects are informed by three main theories—constructivism (learning by doing), constructionism (learning by building), and social constructivism (learning through collaboration and communication) (Amineh and Asl 2015)—all of which complement Knowles' theory of andragogy (Weegar and Pacis 2012).

Thus, the STEM Education course aimed to immerse the teachers in their learning and to actively model elements of what these theories propose: active learning, inquiry-based learning, learner-centred approaches, collaboration and social interaction, reflection and metacognition, multiple representations of knowledge and formative assessment and feedback. For example, to model a student-centred approach, the learning activities purposefully positioned the academic teaching team as facilitators where they helped the learners to explore and develop knowledge through hands-on activities focused on real-world contexts. The foundations developed here promoted critical and creative thinking. Such an environment also promoted collaboration and communication within the classroom. There was a strong emphasis on creating opportunities to ask questions, connect new and existing knowledge, and engage in reflective thinking about their learning processes. Thus, through our in-classroom activities, we were explicitly modelling pedagogies that the teachers could adopt in their own classrooms.

The teaching team contextualised the course readings and workshop activities so that cultural practices and traditions were incorporated where possible. The need for connection to traditional methods is explicitly stated throughout PNG's Mathematics, Science, and Design Technology curricula. For example, in Lower Secondary Science (PNGDOE 2006c), students are expected to: *Compare traditional and contemporary science methods to generate and verify knowledge and explain natural phenomena* (p. 20). In the Lower Secondary Mathematics syllabus (PNGDOE 2006b), students are expected to *Demonstrate an understanding of traditional and contemporary mathematics in Papua New Guinea* (p. 14). Similarly, there is a unit on Village Technologies in the Lower Secondary Design and Technology syllabus (PNGDOE 2006a). As part of their assessment, students are expected to *Use the design process to make at least two simple products or one complex product from a village technology* (p. 32).

In addition, the teaching team was also mindful of some of the other challenges that teachers faced in their classrooms. Lack of resources, large class sizes, quality of learning spaces, availability of digital technologies, internet connectivity, and access to electricity were some of the issues that influenced how the teachers delivered their lessons (Chandra 2019). All these factors were considered, and where possible, the activities within the subjects were designed so that the impact of some of these factors was either minimised or negated. For example, in the fourth subject of the course, *STEM Pedagogies,* the teachers were expected to design and build a boat. Through this task, they were expected to understand the design process and how products are created to meet local needs (Design & Technology). Once the boat was built, they tested it by adding small loads. Through this process, they developed their knowledge and understanding of why boats float and sink (Science). They developed their ideas of mathematical modelling by determining the cost of building such a boat in the real world. The teachers were expected to use any readily available materials rather than sourcing new products from hardware stores. All teachers in the PNG cohort were comfortable with the activity because boat building is a traditional practice where the building materials are sourced locally (Fig. 2.2). This activity would have been achievable in their classrooms because there was little or no cost involved other than the teacher's time. There was no need for digital technologies, electricity, or specialist learning spaces. The activity was achievable in the teachers' context, and they would be able to tailor the activity to their needs.

The simple enjoyable task of boat building embeds a range of STEM skills. For us, it was a symbol of the philosophy on which we had built the whole course. There was no one right way to build a boat. The teachers used their discipline knowledge as well as incorporating Indigenous knowledge in sourcing the right tree for wood and adopting traditional skills in the art of carving their boat with manual tools. The STEM knowledge involved was made explicit to enhance the learning experience. There was no one right pedagogy and the teachers could devise ways of introducing the task in a number of creative ways.

2.2 Summary of Philosophy and Background

This chapter has presented the theoretical foundations and practical considerations that informed the design of the Graduate Certificate in STEM Education course for in-service teachers from Papua New Guinea. While nations worldwide have pushed STEM agendas, the need for trained teachers who can integrate STEM disciplines remains challenging. There is a need for a paradigm shift from the traditional, siloed approaches to teaching science, technology, maths and engineering. For this shift to materialise, teachers need to be upskilled. Research suggests structured professional learning opportunities are

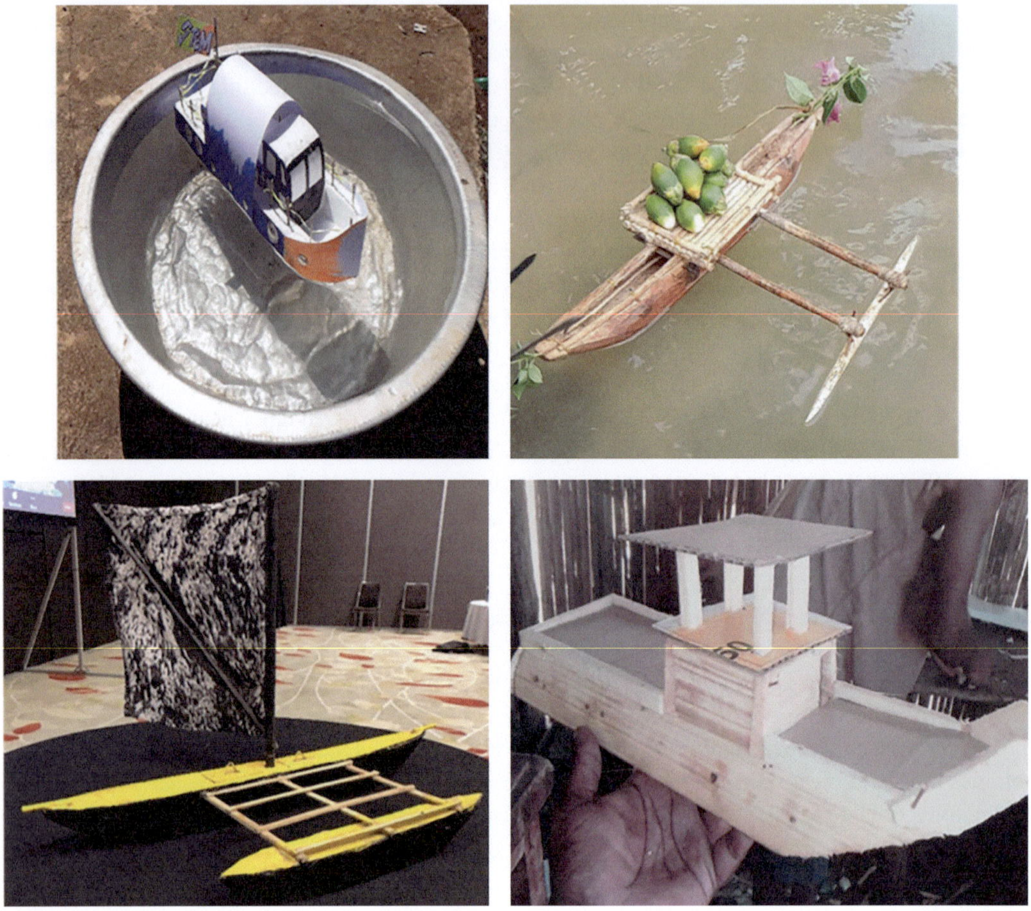

Fig. 2.2 Examples of boats created by the teachers

far more beneficial than conventional and random workshops. To advance the integrated STEM agenda, teachers must be immersed in experiential and contextually relevant training grounded in constructivist and inquiry-based pedagogies that draw upon their disciplinary knowledge and skills.

The PNG teachers in this project enrolled in a context-specific postgraduate course tailored to their needs. The course integrated Bruner's spiral curriculum and Kolb's experiential learning theories, which promote learning through cycles of experience, reflection, conceptualisation, and experimentation. Authentic learning activities strongly underpinned all subjects within this course through pedagogies such as project-based learning, which promoted thinking skills (such as design thinking) and, in the process, facilitated teachers' knowledge and skills to integrate concepts from Science, Technology, Engineering, and Mathematics for classroom activities.

The course consisted of eight subjects, divided into two phases: the first focused on developing pedagogical content knowledge (PCK), and the latter on applying that knowledge in real-world contexts. To support this, a cyclical experiential learning model was employed, beginning with teachers' prior experiences and progressing through classroom application and reflection. A significant strength of the programme was its cultural relevance, aligning course content with PNG's educational priorities and resource constraints. Our example was the boat-building project, which blended Indigenous knowledge and

STEM concepts without reliance on advanced infrastructure. It also introduced the motivational element of fun to teaching and learning.

References

Amineh RJ, Asl HD (2015) Review of constructivism and social constructivism. J Soc Sci, Lit Lang 1(1):9–16

Biesta G (2019) Reclaiming teaching for teacher education: Towards a spiral curriculum. Beijing Int Rev Educ 1(2–3):259–272. https://doi.org/10.1163/25902539-00102015

Bruner J (1960) The process of education. Harvard University Press

Carlson J, Daehler KR, Alonza AC, Barendsen E, Berry A, Borowski A, Carpendale J, Kam Ho Chan K, Cooper R, Friedrichsen P, Gess-Newsome J, Ineke H-R, Hume A, Kirschner S, Liepertz S, Loughran J, Mavhunga E, Neumann K, Nilsson P, Park S, Rollnick M, Sickel A, Kyung Suh J, Schneider R, Jan van Driel C (2019) The refined consensus model of pedagogical content knowledge in science education. In: Hume A, Cooper, R, Borowski, A (eds) Repositioning pedagogical content knowledge in teachers' knowledge for teaching science (pp 79–92). Springer Singapore. https://doi.org/10.1007/978-981-13-5898-2_2

Chandra V (2019) Share engage educate—SEEding change for a better world. Brill

Ejiwale JA (2013) Barriers to successful implementation of STEM education. J Educ Learn (EduLearn) 7(2):63–74. https://doi.org/10.11591/edulearn.v7i2.220

Fink LD (2003) A self-directed guide to designing courses for significant learning. Jossey-Bass

Gess-Newsome JA, Taylor JA, Carlson J, Gardner AL, Wilson CD, Stuhlsatz MAM (2019) Teacher pedagogical content knowledge, practice, and student achievement. Int J Sci Educ 41(7):944–963. https://doi.org/10.1080/09500693.2016.1265158

Gorozidis G, Papaioannou AG (2014) Teachers' motivation to participate in training and to implement innovations. Teach Teach Educ 39:1–11. https://doi.org/10.1016/j.tate.2013.12.001

Guzey SS, Caskurlu S, Kozan K (2020) Integrated STEM pedagogies and student learning. In: Handbook of research on STEM education (pp 65–75). Routledge

Hardman F, Abd-Kadir J, Tibuhinda A (2012) Reforming teacher education in Tanzania. Int J Educ Dev 32(6):826–834

Jorgensen R, Larkin K (2018) STEM education in the junior secondary: The state of play. Springer. https://doi.org/10.1007/978-981-10-5448-8

Katehi L, Pearson G, Feder M (2009) Engineering in K-12 education: understanding the status and improving the prospects. National Academies Press

Kolb DA (1984) Experiential learning: experience as the source of learning and development, vol. 1. Prentice-Hall

Kolb AY, Kolb DA (2009) Experiential learning theory: a dynamic, holistic approach to management learning, education and development. In: Armstrong SJ, Fukami CV (eds) The Sage handbook of management learning, education and development (pp 42–68). Sage. https://doi.org/10.4135/9780857021038.n3

Li Y, Schoenfeld AH, diSessa AA, Graesser AC, Benson LC, English L, Duschl RA (2019) Design and design thinking in STEM education. J STEM Educ Res 2:93–104. https://doi.org/10.1007/s41979-019-00020-z

Loughran J, Berry A, Mulhall P (2012) Understanding and developing science teachers' pedagogical content knowledge, vol. 12. Springer Science & Business Media

Love TS, Hughes AJ (2022) Engineering pedagogical content knowledge: examining correlations with formal and informal preparation experiences. Int J STEM Educ 9(1):1–20. https://doi.org/10.1186/s40594-022-00345-z

PNGDOE (Papua New Guinea Department of Education) (2006a) *Design and Technology Subject Field Lower Secondary Syllabuses.* https://educationpng.gov.pg/Teachers/secondary/lower/syllabus-lower-secondary-design-technology.pdf

PNGDOE (Papua New Guinea Department of Education) (2006b) Mathematics lower secondary syllabus. https://educationpng.gov.pg/Teachers/secondary/lower/syllabus-lower-secondary-mathematics.pdf

PNGDOE (Papua New Guinea Department of Education) (2006c) Science lower secondary syllabus. https://educationpng.gov.pg/Teachers/secondary/lower/syllabus-lower-secondary-science.pdf

Salinger G, Zuga K (2009) Background and history of the STEM movement. In International Technology Education Association. (ITEA), The overlooked STEM imperatives: technology and engineering, K–12 education (pp 4–8). https://www.uastem.com/wp-content/uploads/2019/02/The-Overlooked-STEM-Imperatives.pdf

Shernoff DJ, Sinha S, Bressler DM, Ginsburg L (2017) Assessing teacher education and professional development needs for the implementation of integrated approaches to STEM education. Int J STEM Educ 4:1–16. https://doi.org/10.1186/s40594-017-0068-1

Shulman LS (1986) Those who understand: knowledge growth in teaching. Educ Res 15(2):4–14. https://doi.org/10.2307/1175860

Shulman LS (1987) Knowledge and Teaching: Foundations of the New Reform. Harv Educ Rev 57:1–22. https://doi.org/10.17763/haer.57.1.j463w79r56455411

ThibautL, Ceuppens S, De Loof H, De Meester J, Goovaerts L, Struyf A, Boeve-de Pauw J, Dehaene W, Deprez J, De Cock M, Hellinckx L, Knipprath H, Langie G,Struyven K, Van de Velde D, Van Petegem P, Depaepe F (2018) Integrated STEM education: a systematic review of instructional practices in secondary education. Eur J STEM Educ 3(1):02. https://doi.org/10.20897/ejsteme/85525

Weegar MA, Pacis D (2012) A comparison of two theories of learning—behaviorism and constructivism as applied to face-to-face and online learning. https://www.semanticscholar.org/paper/A-Comparison-of-Two-Theories-of-Learning-and-ivism-Weegar-Pacis/2cb0b3019f04cfbf790d0ed8fa39f603bcfa4f7e

Vinesh Chandra is an Associate Professor and the STEM Subject Area Co-ordinator at Queensland University of Technology (QUT). In 2024, he was awarded an Australian Award for University Teaching (AAUT) for his contributions to advancing STEM education in Australia and overseas. He also served as one of the leaders for the *PNG STEM Education Teacher In-service Project*. For this project, the team was awarded a Gold in the QS Wharton Awards and a High Commendation in the IEAA awards. In addition to being co-editor of this book, Associate Professor Chandra authored this chapter and contributed to others. He has also co-authored other books on STEM education, including the award-winning "STEM Education in the Primary School: A Teacher's Toolkit" published by Cambridge, and "My STEM Workbook" for primary school students.

Course Design and Its Impact on Teachers 3

Vinesh Chandra, Dann Mallet, Senka Henderson, and Matthew Readette

Abstract

The Graduate Certificate in STEM Education course offered to teachers in Papua New Guinea was unique, likely being the first of its kind for teachers in the country. Therefore, understanding the course's impact on the participating teachers was important. This chapter reports on research that specifically focused on the following research question: How did the course design influence teachers' knowledge of integrated STEM education? A mixed-methods approach was adopted, with data collected through a survey and interviews. The research demonstrated that teachers' motivation, continuous concrete experiences and opportunities for reflection, and the conceptualisation of ideas into action plans were the key factors contributing to success.

Keywords

Course design · Mixed methods · Motivation · Concrete experiences · Reflection · Action plans

V. Chandra (✉)
School of Education, Queensland University of Technology, Kelvin Grove, QLD, Australia
e-mail: v.chandra@qut.edu.au

D. Mallet · S. Henderson · M. Readette
Queensland University of Technology, Kelvin Grove, QLD, Australia
e-mail: dg.mallet@qut.edu.au

S. Henderson
e-mail: s10.henderson@qut.edu.au

M. Readette
e-mail: Matthew.readette@qut.edu.au

The Graduate Certificate in STEM Education at Queensland University of Technology (2020–2023) and its participating teachers from Papua New Guinea are special. The focus of the qualification, that is, STEM Education, calls for a rethinking of pedagogy and curriculum planning which, when located in a developing country keen to improve its education system and kickstart its economy, warrants closer examination.

This need for examination gained urgency when we found ourselves, as an academic teaching team, forced by circumstance into a situation of rapid redesign as our first cohort of students were forced away from the university and out of the country as COVID-19 made its presence felt. We knew immediately that this situation, given that our students lived and worked mostly in rural communities, was worthy of investigation. It became more pressing when we found that there was little research on multimodal models of

V. Chandra et al. (eds.), *Integrated STEM Education in Papua New Guinea*, Sustainable Development Goals Series, https://doi.org/10.1007/978-981-95-0828-0_3

learning undertaken by in-service teachers from developing countries.

Most extant studies focus on participants learning in multimodal environments in higher education institutions in developed countries (e.g. Castle and McGuire 2010; Konstantinidou and Nisiforou 2022) often finding that learning outcomes in online courses were the "same or superior to those in face-to-face" learning (Allen and Seaman 2013, p. 5). We wondered if the teachers in our cohorts would have the same experience. We further wondered if our course design (see this chapter) would be robust enough to weather modes of delivery other than face-to-face workshops. We also wondered if online learning would work for the teachers as they wrangled the myriad infrastructure and resource problems they could potentially face back home in PNG.

To this end, we initiated research in parallel with the conduct of the course. As noted, we posed two research questions:

- How did the course design impact teachers' knowledge of integrated STEM Education?
- How did the modes of course delivery impact on teachers' learning?

This chapter deocuments the findings of the first research question on course design described in the previous chapter (Chap. 2). Our response to the second research question is presented in the following chapter (Chap. 4).

3.1 Methodology

Our research project adopted a qualitative case study approach. Qualitative case study research designs have been used in similar studies. For example, Han et al. (2015) investigated the impact of professional development sessions (10 sessions per year, seven hours per session over a three-year period) on 92 high school teachers in the USA. The study specifically targeted the teachers' capabilities of applying their knowledge and skills of STEM Project-Based Learning (STEM-PBL) in their classrooms. According to Creswell (2014), case studies enable researchers to develop "an in-depth analysis of the case" which is often bound by "time and activity, and researchers collect detailed information using a variety of data collection procedures over a sustained period of time" (p. 14). Case studies are also appropriate where theories are not readily available to investigate participants' behaviour (Yin 2003).

A case-study approach suited our overarching intentions of investigating the impact of the postgraduate certificate course in STEM Education on the PNG teachers. In this instance, the PNG teachers are the case, with unique qualities derived from their country and classroom contexts being substantively different from most Australian classrooms and those in other developed countries. The data was gathered through a variety of procedures including interviews and a survey. A qualitative case study research design enabled us to focus in-depth on the PNG teachers and how the course design and delivery impacted on them. This research provided us with rich, detailed, and contextual insights into the complexities of undertaking a professional learning course in STEM Education from the PNG teachers' perspectives.

3.1.1 Research Participants

All 75 enrolled teachers (36 males and 39 females) across the three cohorts were invited to participate in this research via an online survey. Sixty-four survey responses were received (Teacher 1–Teacher 64).

Further (and inclusive) to this, 12 participants from across the three cohorts took part in semi-structured interviews. Within Cohorts 1 and 2, eight teachers (4 males and 4 females) volunteered to participate in two semi-structured online interviews lasting 30–40 min. Within Cohort 3, four teachers (2 males and 2 females) participated in the second interview. A summary of the participating teachers (with pseudonyms) is presented in Table 3.1.

Table 3.1 Summary of interview participants ($N = 12$)

Teacher	Cohort	Gender	Teaching Experience	Location (Urban, Rural)
Elizabeth	1	Female	15+ years	Urban
Michael	1	Male	10+ years	Urban
Rose	1	Female	15+ years	Rural
William	1	Male	15+ years	Urban
Miriam	1	Female	10+ years	Rural
Joan	2	Female	15+ years	Urban
Gabriel	2	Male	10+ years	Rural
Joseph	2	Male	10+ years	Urban
Walter	3	Male	10+ years	Rural
Sarah	3	Female	15+ years	Urban
Amos	3	Male	15+ years	Rural
Lucy	3	Female	15+ years	Rural

3.2 Data Collection

Our qualitative research made use of two data instruments: a survey and semi-structured interviews. They were conducted over time and when convenient to research participants.

3.2.1 Survey

The survey was administered online once the teachers from each cohort had graduated from the course. We anticipated that the survey would take 30 minutes to complete. Given the challenges of access to technology, connectivity, and other commitments, we allowed up to a month for the teachers to respond. Across the three cohorts, the survey response rate was exceptionally high with 64 teachers (85%) completing the survey. The survey gathered demographic data through categorical response questions such as: "How long have you been teaching?" and "What are your teaching qualifications?" Open-ended survey questions gave us an idea of the participants' home and school circumstances as well as their perception of the STEM Education course itself. Examples of open-ended questions included:

- Can you identify three new ideas that you learnt about integrated STEM Education through your enrolment in this course?
- How do you intend to apply the new ideas in your classroom or in your school or in your context?
- What challenges and barriers do you foresee to apply your ideas in your classrooms?
- How can the challenges and barriers be overcome?
- For a short duration, some of you were able to participate in the course through a face-to-face mode. For most of the time, you participated in an online mode either from your school/home or from a hotel in Port Moresby.

  - In your view, how did the different modes impact on your needs as a learner?
  - How did the modes impact on your learning?

The survey gave us a global picture of how the design and the delivery of the course impacted on the teachers across the three cohorts. How the questions relate to the two research questions is self-evident, particularly in the direct request for a comparison between the different modes of learning and teaching we adopted.

3.2.2 Interviews

The semi-structured interviews were designed to allow for in-depth questioning by providing key probes to the set questions (Creswell 2014). The interview questions probed the participants' learning experiences, as well as the issues and challenges that influenced their engagement in terms of course delivery modes. For instance, participants were asked about the challenges they faced when switching between online and face-to-face modes of learning, and which mode they considered more effective in meeting their needs as learners. The interviews were conducted in the participants' own time and were audio- and video-recorded via Zoom. Despite occasional connectivity challenges that arose while conducting the interviews from teachers' homes or schools, we managed to overcome these issues and completed the interviews with teachers. In all, a total of 12 teachers (6 males, 6 females) from across the three cohorts participated in the 30-min semi-structured individual interviews (see Table 3.1). Coincidentally, the teachers represented a range of locations and teaching experience. The interviewees were asked questions such as:

- You experienced different modes of course delivery? How did these modes impact you?
- What were some of the strengths of these modes?
- What were some of the weaknesses/challenges and how did you deal with them?
- How did the design and delivery of the course impact on the different modes? What were some of the highs and lows?
- Was the design and delivery of the Graduate Certificate in STEM Education contextually appropriate? Please explain your answer.

3.3 Data Analysis

The online survey tool summarised the responses to the short categorical response questions. The qualitative data from the survey was analysed manually using the steps proposed by Braun and Clarke (2006). We read and re-read the survey responses to familiarise ourselves with the teachers' responses and in the process generated the initial codes for the data. From the codes we developed and consolidated the themes which provided insights into the teachers' responses. Additionally, the qualitative data analysis software, NVivo, was used to explore for the emergence of any additional themes.

The survey data was loaded directly using the Qualtrics integration which loaded the survey as cases eliciting case attributes describing each of the cases produced from the close-ended questions. The open-ended question responses were grouped in nodes to allow for pattern and theme exploration. The interview data was loaded in two ways. The first was with each participant response as a case, the second approach was collating all responses to each question with each question serving as a case. Analysis could then occur across each participant's case to see if there were themes emerging from the individuals and then across all responses to see if themes were emerging from the cohort. In the following section, we focus on the impact on the teachers of the *design* of the postgraduate course in STEM Education, that is, we respond to our first research question.

3.4 Findings

In the previous chapter (Chap. 2), we described how we had designed the Graduate Certificate in STEM Education for three successive cohorts of teachers from PNG. To begin with, the survey data unequivocally pointed to overall satisfaction with the course (see Table 3.2). We asked for a rating against the nine attributes of course design (drawn from Fink 2003). The teachers ($n = 64$) rated the course highly across all nine attributes in the survey, responding on a five-point scale (1 = Unsatisfactory, 2 = Satisfactory, 3 = Good, 4 = Very Good, and 5 = Excellent).

Table 3.2 Course satisfaction results (drawn from Fink 2003)

Attributes	Model of course design	Min	Max	Mean	SD
1. Quality of course content	Teaching/learning activities	2	5	4.52	0.70
2. Relevance of learning resources	Teaching/learning activities	2	5	4.36	0.69
3. Relevance of course to PNG curriculum	Connection to context	3	5	4.27	0.71
4. Relevance, of course to your context	Connection to context	2	5	4.00	0.74
5. Knowledge and expertise of lecturers	Teaching/learning activities	3	5	4.70	0.58
6. Quality of assessment tasks	Feedback and assessment	3	5	4.48	0.66
7. Quality of digital technologies	Teaching/learning activities	2	5	4.36	0.77
8. Effectiveness of the Learning Management System (LMS) interface	Teaching/learning activities	2	5	4.42	0.74
9. Overall quality of the course	All	2	5	4.48	0.66

3.4.1 Motivation to Study as Factor in Course Satisfaction

To attribute teachers' positive satisfaction levels to our course design, we believed it was essential to understand their expectations in terms of their reasons for enrolling in the course. In this way, we could ascertain if (and perhaps how) their expectations were being met.

Teachers' responses to the open question, *why did you enrol in the Graduate Certificate in STEM Education,* were quite compelling. Interestingly, none of the teachers enrolled in the course to develop their content knowledge even though none were subject experts in *all* four STEM disciplines. They all wanted to either build their skills in teaching Mathematics or Science or acquire knowledge about the recently introduced PNG STEM Education curriculum (PNG PM Marape 2020).

Some reported an ardent desire to enhance the core business of teaching and learning in the classroom with a view to preparing their students for "21st-century challenges". William (interview) believed his students were exposed to *so much information* via the internet and he believed he needed to *upskill* himself *with more specific skills and knowledge.* Many teachers felt that their pedagogical approaches were outdated and needed a makeover. As a teacher, Elizabeth (interview) wanted to be *on par with the rest of the world and see what's happening.* She saw the Graduate Certificate as *a rare opportunity to develop new skills* because in PNG teachers did not have *very many professional development programs, where we are exposed to what's happening in classrooms outside.* They were mindful of the initiatives of the PNG government to introduce STEM Education as a means to improve education, enhance the economy, and ensure the wellbeing of the country.

Others explained their reasons for applying to join the course as follows:

- Realistically, I want to bring change in the lives of my students and make them experience hands-on skills to become problem solvers, critical thinkers, inventors, designers, etc. so that they can handle and deal with real-life problems. [Teacher 2]
- To help improve the boring way of teaching mathematics. [Teacher 16]
- To develop understanding and skills on STEM Education and its implementation. Currently, in PNG, we have a new STEM syllabus offered to six PNG National Schools of Excellence and enrolling in this course is very important in the implementation stages. [Teacher 20]
- I decided to join this course … to look purposefully at other alternatives to make my teaching more effective and suitable for the learners [Gabriel, interview]

Drawing on the findings of Girvan et al. (2016), all recognised that their pedagogical approaches in their classrooms required a significant transformation. They also wanted to learn more about STEM Education itself. According to Gorozidis and Papaioannou (2014), if teachers are "autonomously motivated", they are more likely to engage in professional learning activities (p. 8). However, buy-in into reforms will only occur if it aligns with teachers' beliefs about how teaching and learning should occur (e.g., Avalos 2006). Changing teacher beliefs is not easy. Teachers may be forced to enact reform agendas, but this does not necessarily alter their beliefs and consequently has a limited impact on their classroom practices. Girvan et al. (2016) investigated the effect of an experiential learning professional development program on 38 in-service teachers in Ireland. They reported that, when teachers believe in a reform initiative, "positive narrative(s)" emerge about the reforms (Girvan et al. 2016, p. 138). In the case of the PNG teachers, they all wanted to be in the course. A positive sense of their engagement to support forthcoming reforms in PNG is evident in teachers' responses. For example:

> Next year the curriculum in PNG will change from outcome-based education to Standards Based Education (SBC). About 40% of the curriculum is STEM related. Thus, the education division in our provinces are looking at us to advocate STEM pedagogies to other teachers in the province. [Teacher 13]

Teachers' positive beliefs were beneficial to their genuine participation because they were not forced into enrolling in the course. A positive frame of mind points to intrinsic motivation, which can strongly influence how learners engage and interact with their learning. In this case, all teachers were quite focussed on bringing about a change in their classrooms and at the same time developing their knowledge in STEM Education. None of the participants indicated that they enrolled because it was offered by an Australian university and they had a chance to travel overseas. Further, no teacher had been asked (or required) to enrol in the course by their employers. The high level of satisfaction reported in the survey data would indicate that their positive expectations had been met by the course and that they were buoyed by their autonomous motivation to participate.

3.4.2 Ongoing Concrete Experiences and Reflection Opportunities as Factor in Satisfaction

As previously noted (Chap. 2), the course was initially designed to be delivered in discrete blocks at the university in Australia interspersed with periods of full-time work in the classroom by the teachers. This model enabled the teachers to continuously apply the knowledge and skills gained in the different subjects, even before graduation from the course. We had developed this schedule from Phase 2 of our cyclical experiential learning model (Chap. 2, Fig. 2.1).

There was a three-way interaction between the teachers' concrete experiences in the course, concrete experiences of their new knowledge, and skills in their school classrooms and their personal reflections (Fig. 3.1).

We hoped that this pattern of enforced reflection and application would encourage teachers to make measured but permanent changes to their pedagogy. We saw ample evidence of how teachers applied their new knowledge in their classrooms. For example, teachers reported physical modelling (viz., 3D clay-models of an ionic compound, a model of human anatomy) and project-based learning (viz., making a coconut scraper, production of electrical circuit boards and accessories) in place of lecture-style teaching. Some also adopted digital technologies to facilitate students' learning. These reported approaches connected with concepts present in the course subjects. Teacher 20 reported that:

> In Grade 11 Physics, a project on projectile motion was carried out in an open space. Students were told to find the time it takes for a soccer ball to be in flight, measure the distance (horizontal) the ball in flight travelled and calculate the angle at which the ball leaves the striking foot. They captured images and videos of their practicals. The project was submitted in PDF form with pictures included.

The approach adopted by Teacher 20 was a "game changer". Traditionally, students would be taught about projectile motion through "chalk and talk", equations, and graphs, with the teacher in control. However, getting the students to play with a soccer ball to develop conceptual understanding

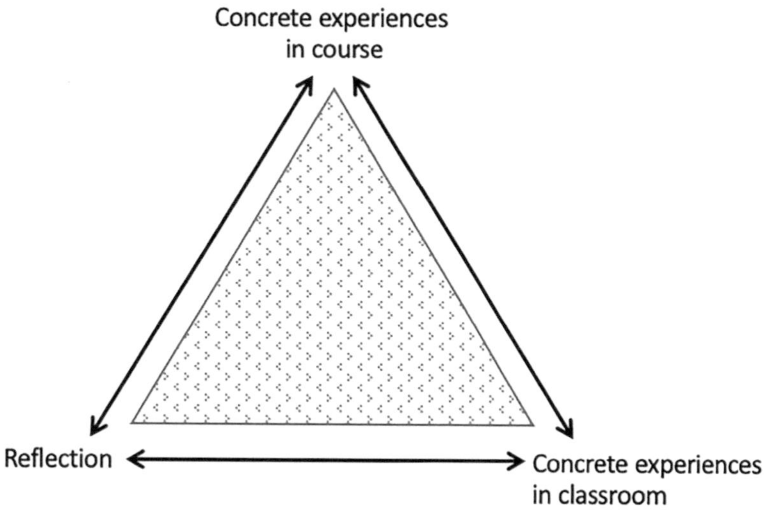

Fig. 3.1 Interaction of experiences and reflection

of time and motion immerses them in the activity as active learners.

Similar shifts in pedagogy were highlighted in teachers' responses because they believed their *teacher-facilitated, student-centred learning* activities promoted *collaboration* because the tasks were *authentic,* and students could relate to what they were doing. Well-designed activities with well-thought-out rubrics also facilitated the teaching and learning process:

> The apparent difference from what I've been used to is that students collaborated and came up with a prototype that worked. The feeling you get at this point is truly satisfying and fulfilling. I did less of the traditional teaching and more facilitation and mingling. Most materials used were from garages, dumps and the local environment. [Teacher 31]

> Students were given the topic and told to research and propose methods and ways to construct a biogas model. Then, every piece of activity is assessed using specific rubrics. [Teacher 9]

The teachers could sense that this shift in their pedagogical approach was making a difference because the students *owned what they had created—they even invited their friends to see what they had done* [Teacher 33]. In addition, *there was excitement and curiosity* [Teacher 59] and a *passion* [Teacher 31] for working on the project. For reasons such as this, student engagement was high in all activities: *regardless of their IQs, everyone participated because the activity guided the students towards the intended outcomes* [Teacher 45].

Teacher 35 believed that a three-week class activity on measurement was *very engaging.* Students learnt how to use different measuring instruments such as trundle wheels, tape measures, and clinometers. Teacher 35 reported that:

> One student brought a drone to measure the height of the taller buildings as a verification tool and view the building from several vantage points. The students also discovered various apps on their phones to help measure the angles and heights of the buildings.

Following their in-course experiences, teachers were becoming bolder in terms of how they were showcasing their students' work. Joan (interview) noted that she was already *practising* what she was learning, pointing out that *I have projects which they are in my YouTube channel.* Teacher 35 also noted that:

> Several students did not do any practicals in their Design and Technology classes and appreciated the exercise; they became the leaders in their groups, teaching their peers the different symbols to use on their 2D plans to represent the features in the classrooms. Learning outcomes were achieved by applying both Mathematical and Design and Technology concepts. Both staff and students gave positive feedback in their evaluations and recommended that we in the Mathematics department do more tasks/projects. Hence, the students can appreciate and see the applicability of what is taught in the classroom.

William (interview) offered that the course was giving him new ideas on educational frameworks and how they can be implemented. Prior to starting the course throughout his 10+ years of teaching, he would just go *in the classroom and start* the lesson anywhere. He would be *jumping between ideas* in *the lesson* and, as long as the content was delivered, he was satisfied. However, since *I started to integrate STEM Education … I learnt that you always* need a structure such as starting by *identifying the problem.* The explicit connection between theory and practice was clear and enhanced by the scheduling of coursework and classroom time. The teachers had become aware of what they were learning and its immediate value to them. There were deliberate and welcome changes occurring in teaching practice directly due to the course content.

3.4.3 Conceptualisation of Ideas into Action Plans

Our cyclical experiential learning model (see Chap. 2) enabled the teachers to progressively deposit knowledge and progress in conceptualising how different pedagogical approaches are adopted to deliver integrated STEM. They felt confident with hands-on activities with real-world connections which develop students' problem-solving and systems-thinking skills. Project-based learning and the 5E Instructional Models (Bybee 2010; Forbes et al. 2021) were predominantly identified as the models that gave them strong frameworks for STEM integration.

The teachers also believed that the course had developed their digital literacy and fluency. They felt more competent applying their new knowledge and skills whenever they had access to digital technologies. They also developed current ideas for designing *assessment tasks* and accompanying *rubrics*. The teachers also indicated that they had learnt new ways of doing activities by connecting with experts within their communities. Teacher 22 listed what had been learnt which had not, for them, been previously apparent:

- Systems Thinking (For school principals, this is very vital); and,
- Realising that Mathematics is not a theoretically oriented subject, it has many real-world applications.

Similarly, Teacher 32 offered the following as new understandings:

- I, as the teacher, am only a facilitator;
- Learners will learn better if they investigate by themselves;
- Resources in learning are all around me to use in my teaching every day; and,
- Allow learners to be change-makers and problem solvers.

Many pointed out that what they were learning was contextually appropriate. Joan (interview) explained that the STEM Education course was enabling her to connect her *planning, teaching and assessment* more effectively as per the PNG Standards-Based Curriculum. She was also finding the use of technologies to be remarkably effective. She was so confident of her strategies that she *conducted a few in-house in-service sessions for teachers* to develop their skills as well. With Joan's guidance, *some teachers were also using technology to teach and had given me feedback that technology makes it very easy for them to teach the content.*

In the final subject, *Applied STEM in Education* (see Chap. 2, Table 2.1), the teachers were asked to submit an action plan in the form of a book chapter (a selection of which was selected for inclusion in Part II of this book). In their action plans, teachers had to outline how they intended to apply their knowledge and skills within their contexts. The rationale of this task was to enable the teachers to think as leaders in STEM Education and consider how they intended to play their roles in their educational settings. Thus, the action plan/book chapter needed to present an integrated STEM idea they were undertaking or intended to undertake. Through the plan, the teachers were expected to articulate "the why, the how, and the what" of their proposed ideas. A required rationale

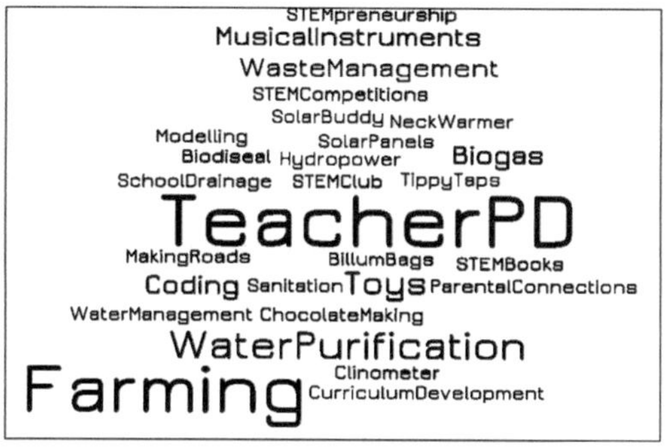

Fig. 3.2 Teacher project themes

needed to be evidence-based. Understandably, the teachers had a range of different ideas. Rose (interview) pointed out that this subject enabled them to *contextualize* [the content], *it was easier for us to relate back.* The most popular themes for the projects and ideas were teacher professional development, farming and water purification (Fig. 3.2). Other project ideas included creation of toys, sanitation, STEM clubs, and chocolate making. All ideas had a connection with at least one Sustainable Development Goal (UN., n.d.).

Most of the teachers in the Graduate Certificate course felt that upskilling their colleagues was vital. They reasoned that while they could implement an integrated approach in their classrooms, the whole school needed to be competent in taking on this approach with confidence. Unless this was done, the country's integrated STEM agenda was unlikely to succeed (see Chap. 9 for narratives relating to teacher leadership in designing and delivering professional development).

Access to clean water is also a significant problem in PNG. Thus, several teachers took on the challenge of designing and implementing class projects where their students' proposed solutions on how clean water could be made available to the school community (see, e.g., Narratives 7.4, 8.1). Further, teaching students about farming through STEM was considered a priority by some of the teachers (see, e.g., Narratives 8.4, 9.3). The reasoning here was that, for many students, a tertiary pathway was not a realistic option. They believed that if students were taught about effective farming practices through the application of STEM, then they would be better equipped to fend for themselves in real-world post-secondary school education.

Despite viable plans, many teachers felt they would have to overcome various contextual barriers before implementing their ideas. Teacher 61 explained that *the challenge is managing and motivating a large class.* A range of other factors was also identified, such as: school administrators not backing teachers' ideas; contemporaries not wanting to accept change; school timetabling; lack of resources; lack of community support; classrooms and furniture built on traditional learning styles; and frequent power blackouts. Teacher 31 outlined numerous barriers to the integration of STEM as follows:

- Lack of appropriate laboratory equipment has barred me from running experiential learning in class over the years. I have resorted to planning classes that would use trash materials instead.
- Even recycled materials must be handled with proper technological tools to yield an expected outcome. I am acquiring some in the lab, such as soldering rods, glue guns, soldering wire and conducting wires, but I have asked students to source other resources independently.
- The internet connectivity in the school is unreliable. I pay for data (with my own money) and have been accessing it via my phone to date.
- I wish to access a projector and scanner in my office to be used in class preparations and presentations without inconvenience and the protocols to meet in accessing the school's projector.
- The class size of 40 is too much for the lab, which can hold only 30 students. I prefer 30 students per class who can sit comfortably in the allocated seats.

Despite these challenges, there was optimism. The teachers all had some concrete ideas on how they could move forward with their plans. For example, Teacher 31 proposed the following strategies to deal with the listed challenges:

- I'm planning projects enabling students to use local resources, even trash.
- The school administration must come out of its arrogance and purchase lab equipment with budgeted funds not diverted elsewhere. The diversions over a decade have ripped the fun and experiences in science that the subject deserves.
- I have decided to buy equipment yearly with the money allocated and build the department stocks over time. I started last year.
- I have created a class WhatsApp group where resources and notes are sent for student access. I also use my laptops to deliberate during individual consultations.
- I occasionally take students off class on the grass to put everyone on the same level. I have put in place extra chairs in the class for students. I still wish for a 30-class size.

To tackle the issue of the lack of support from school administrators, fellow teachers, parents, and the school community, Teachers 14 and 35 proposed the following approaches:

> I need to increase awareness and educate teachers, parents, and communities to fully understand the new concept of STEM before implementing the STEM programs in the school. Then, all I need to do is to capture their interest in the STEM agenda, and everything else should flow smoothly. [Teacher 14]

> More professional development time and the formation of a STEM committee or PLCs to allow for follow-ups and continuity of the program. Without these, there won't be any continued dialogue between the team and the appreciation of what integrating STEM offers to the school. [Teacher 35]

Teacher 43 decided that he would demonstrate his idea in his class before sharing with others. *I will start small (take baby steps) and accept whatever is given or available. Then, I'll use it and create avenues where it will be possible for others to see the fruits of what I'm doing.* This strategy can be effective because the STEM Education graduates, with their ideas, are stepping in as innovators. Changing classroom practices which have existed for decades cannot be changed overnight. These practices are not light switches that can be turned on or off. According to Conyers and Wilson (2016), one of the ways to gain support for positive change in an educational setting is by clearly communicating the observable benefits that can be demonstrated through practice.

3.5 Summary of Course Design and Impact

This chapter has described our research methodology and introduced the research questions. It has presented the findings from the first research question which addressed the impact of the course design on the teachers. We found that there was a high level of satisfaction with the course which we believed meant that teachers' expectations were being met. One teacher offered that *the course got me on my feet* [Teacher 24].

We found that our planned block schedule of classwork and time for reflection and application was resulting in teachers making deliberate changes to their practice. This went hand in hand with a growing confidence in themselves and acceptance of the STEM pedagogies we were promoting. Lastly, we found that teachers saw beyond themselves as change agents and became actively concerned with their colleagues and school structures. Despite their cumulative experience in the classroom: 5–10 years (26%); 11–15 years (46%); 15 + years (28%) (see Introduction to Part I, all were open to change. They saw STEM Education as contributing to whole communities through engagement in authentic problems and began to plan projects and activities to improve utilities such as water and electricity. They embraced sustainability and saw value in simple critical activities such as teaching students about farming and recycling. We believe that this provides evidence of the success of our course design in contextualising STEM Education in PNG.

References

Allen IE, Seaman J (2013) Changing course: ten years of tracking online education in the United States. Babson Park, MA: Babson Survey Research Group and

Quahog Research Group. http://www.onlinelearningsurvey.com/reports/changingcourse.pdf
Avalos B (2006) The curriculum and professional development of teachers. Prelac J 3:104–111
Braun V, Clarke V (2006) Using thematic analysis in psychology. Qual Res Psychol 3(2):77–101. https://doi.org/10.1191/1478088706qp063oa
Bybee RW (2010) Advancing STEM education: a 2020 vision. Technol Eng Teach 13(6):30–35
Castle SR, McGuire CJ (2010) An analysis of student self-assessment of online, blended and face-to-face learning environments: implications for sustainable education delivery. Int Educ Stud 3(3):36–40. https://doi.org/10.5539/ies.v3n3p36
Conyers M, Wilson D (2016) Smarter teacher leadership: neuroscience and the power of purposeful collaboration. Teachers College Press.
Creswell JH (2014) Educational research: Planning, conducting, and evaluating quantitative and qualitative research, 4th ed. Pearson
Ejiwale, J. A. (2013). Barriers to successful implementation of STEM education. J Educ Learn (EduLearn) 7(2):63–74. https://doi.org/10.11591/edulearn.v7i2.220
Fink LD (2003) A self-directed guide to designing courses for significant learning. Jossey-Bass
Forbes A, Chandra V, Pfeiffer L, Sheffield R (2021) STEM education in the primary school: a teacher's toolkit. Cambridge University Press
Girvan C, Conneely C, Tangney B (2016) Extending experiential learning in teacher professional development. Teach Teach Educ 58:129–139. https://doi.org/10.1016/j.tate.2016.04.009
Gorozidis G, Papaioannou AG (2014) Teachers' motivation to participate in training and to implement innovations. Teach Teach Educ 39:1–11. https://doi.org/10.1016/j.tate.2013.12.001
Han S, Yalvac B, Capraro MM, Capraro RM (2015) In-service teachers' implementation and understanding of STEM project-based learning. Eurasia J Math, Sci Technol Educ 11(1), 63–76. https://doi.org/10.12973/eurasia.2015.1306a
Konstantinidou A, Nisiforou E (2022) Assuring the quality of online learning in higher education: Adaptations in design and implementation. Australas J Educ Technol *38*(4):127–142. https://doi.org/10.14742/ajet.7910
PNG PM Marape launches Schools of Excellence Policy (2020, June 15) Papua New Guinea Today. https://news.pngfacts.com/2020/06/png-pm-marape-launches-schools-of.html
United Nations (UN) (n.d.) Sustainable development goals. https://www.un.org/sustainabledevelopment/sustainable-development-goals
Yin RK (2003) Case study research: design and methods, 3rd ed. Sage

Vinesh Chandra is an Associate Professor and the STEM Subject Area Co-ordinator at Queensland University of Technology (QUT). In 2024, he was awarded an Australian Award for University Teaching (AAUT) for his contributions to advancing STEM education in Australia and overseas. He also served as one of the leaders for the *PNG STEM Education Teacher In-service Project*. For this project, the team was awarded a Gold in the QS Wharton Awards and a High Commendation in the IEAA Awards. In addition to being co-editor of this book, Associate Professor Chandra authored this chapter and contributed to others. He has also co-authored other books on STEM education, including the award-winning "STEM Education in the Primary School: A Teacher's Toolkit" published by Cambridge, and "My STEM Workbook" for primary school students.

Dann Mallet is Professor of STEM Education and Professor of Mathematics at Queensland University of Technology. He is experienced in academic and curriculum team leadership and mentoring, as well as extensive experience as a university Mathematics and STEM educator at all levels. Dann has initiated and led significant international education aid projects in the Indo-Pacific region, and has been recognised through numerous awards and nominations, including, most recently, a Gold in the QS Wharton Awards and High Commendation in the IEAA awards. Aside from his role as co-editor, he has contributed to Chap.s 3 and 4. Along with Vinesh Chandra, he wrote the Introduction to and Review of Chap. 8 *Project-Based Learning*.

Senka Henderson is an Associate Professor at QUT, a Senior Fellow of the Higher Education Academy (HEA) and an Associate Fellow (Indigenous) of the HEA. She is the Course Co-ordinator of the Master of Teaching (Secondary) programme. Senka teaches multiple curriculum, assessment, and pedagogy units in secondary studies across both undergraduate and postgraduate courses. She was involved in the case study research which underpins Part A. She contributed to Chap. 3 and led the writing of Chap. 4. Along with Charisse Farr, she wrote the Introduction to and Review of this chapter *Authentic Learning*.

Matthew Readette is a Research Fellow, sessional academic, and a Fellow of the Higher Education Academy (HEA). He has recently completed his PhD and has taught various units and short courses focussed on teacher education frameworks for global education, as well as academic and digital literacies. His research focuses on social and spatial mobility, international education and teacher education. He was involved in the research underpinning Part A and has made contributions to Chaps. 3 and 4.

4 Course Delivery and Its Impact on Teachers

Senka Henderson, Vinesh Chandra, Matthew Readette, and Dann Mallet

There is no doubt that ongoing professional learning is essential for in-service teachers to effectively address the evolving needs of education systems. The process of designing professional learning courses is distinct from their delivery, and both aspects require careful consideration. In-service teachers often face the challenge of finding time to participate in a formal course at a tertiary institution in addition to fulfilling their regular duties. In some cases, the school may allow teachers time so that they can engage in their professional learning. Committing to professional learning also means time away from families and other day-to-day activities. Another significant point of note is the fact that learning environments are also changing due to the adoption of digital technologies, as these technologies are constantly evolving. A teacher who graduated a decade ago would find the learning environment quite different in a tertiary setting. For reasons such as these, the provider of professional learning courses needs to be aware of and responsive to the most suitable delivery modes for in-service teachers.

Australian Government short courses are traditionally offered in intensive teaching blocks in face-to-face and blended modes (Australian Government n.d.). As previously noted, this was the proposed mode of course delivery for the PNG teachers in the first two cohorts. However, unforeseen circumstances arose due to the COVID-19 pandemic, which was at its peak during that time, travel restrictions, and border closures between PNG and Australia presented new and unprecedented challenges for both the teachers and the university team. This led to some new and unforeseen challenges for all of us. A course originally intended for face-to-face delivery in Australia, heavily reliant on technology and physical resources, needed to be reconceptualised to effectively engage learners online across remote regions of PNG. In contrast to developing countries, the university team faced a multitude of challenges that are extensively discussed in the literature on Papua New Guinea and developing countries (e.g., Aung & Khaing 2016; Chandra 2019; Oko 2021). These challenges encompassed limited access to technology, unreliable connectivity, low digital literacy, inexperience with online learning, inconsistent electricity supply,

S. Henderson (✉) · M. Readette · D. Mallet
Queensland University of Technology, Kelvin Grove, QLD, Australia
e-mail: s10.henderson@qut.edu.au

D. Mallet
e-mail: dg.mallet@qut.edu.au

V. Chandra
School of Education, Queensland University of Technology, Kelvin Grove, QLD, Australia
e-mail: v.chandra@qut.edu.au

V. Chandra et al. (eds.), *Integrated STEM Education in Papua New Guinea*, Sustainable Development Goals Series, https://doi.org/10.1007/978-981-95-0828-0_4

and so on. Understandably, the teaching team in Australia faced a significantly different context from what they were accustomed to. As a result, different modes of course delivery had to be adopted to address these challenges.

The innovative approaches devised to overcome these challenges provided valuable insights that can inform the delivery of distance learning beyond the pandemic era while also enriching the learning experience for students in PNG. In this chapter, we discuss those challenges specific to the learners and context of PNG, which threatened to put an end to an important opportunity for the teachers enrolled in the course. The approaches employed to successfully overcome the challenges have instead resulted in new opportunities for course delivery in developing countries.

In higher education, traditional face-to-face teaching is rapidly becoming outdated and overshadowed by the advancements of digital technologies. Thus, blended and online now dominate teaching and learning. Blended learning combines the traditional in-person instruction with online learning activities allowing institutions to determine the balance between face-to-face and digital components based on learners' specific needs. Blended learning offers the best of both worlds. Online learning, on the other hand, takes the face-to-face component out and all learning activities are delivered through digital platforms and the internet. It has been contended, as previously noted, that online courses are on a par with or superior to face-to-face learning (Allen and Seaman 2013).

Moving from a face-to-face to online learning not only removes the human element of interaction between student to student and student to teacher, but is also dependent upon a reliable and fast broadband connection that can support online learning (OECD 2020a, 2020b). Despite advances in communication technologies, there are geographical areas and underserved communities where this is a major issue. For example, in many OECD countries, "fewer than half of rural households are located in areas where fixed broadband at sufficient speeds is available" (OECD 2020a, p. 3). The 38 countries in the OECD are more prosperous than many countries in the developing world. Thus, if connectivity is an issue in these countries, then citizens in developing countries would have far more issues with connectivity. Affordability and access to technology (e.g., laptop, desktop) is another issue. A decade or so ago, most developed nations encountered and endeavoured to address issues such as affordability, connectivity, and coverage, giving them a first-mover advantage (Zhang et al. 2018). However, these challenges persist in countries like PNG, placing their population including the workforce and teachers and their students at a significant disadvantage. The situation is further compounded by the slow adoption of technology and limited digital literacy and skill development, making international online teaching particularly precarious, especially when coupled with a global pandemic. Suddenly shifting an entire course online exposes the pressing issues that arise in such circumstances.

Apart from technological hardware and software issues, there were other matters of concern at the human level. According to OECD (2020b), a lack of basic digital literacy can have a significant impact on learners' capacities to adapt to a new situation especially if it occurs abruptly. Academic capabilities of learners can also be a factor that can influence student success in online learning (Badrul et al. 2021). The shift from a structured blended course in a university setting to online learning also raises the issue of time to commit to study and engage with the learning materials. Despite some of these challenges, "students' attitudes and dispositions to learning, such as ambition or motivation, are important drivers of their educational achievements and can help ensure that online learning is as effective as possible" (OECD 2020b, p.13).

The literature extensively covers the discussions surrounding the abrupt shifts in course delivery methods caused by the global pandemic, as numerous higher education institutions had to swiftly adapt. Due to the unique circumstances of online teaching during the pandemic, Hodges and Fowler (2020), introduced the term "emergency remote teaching" (ERT) and defined it as "a temporary shift of instructional delivery to an alternate delivery mode due to crisis circumstances" (p. 6). Hodges and Fowler (2020) maintained that the conditions for effective online learning include careful instructional design, planning, and development, as well as an investment in support systems. Certain courses, particularly those in STEM fields, posed challenges when it came to transitioning to online delivery due to the limited availability of in-person interactions. STEM disciplines often rely on hands-on experiences and laboratory work, making it difficult to replicate the same level of engagement and practical learning in an online environment.

The current study explores the conditions of ERT highlighted by Hodges and Fowler (2020). More specifically it focusses on the adjustments to the delivery modalities and the experiences of the PNG teachers along the way. What sets the in-service course in this project apart is the unique journey that was undertaken by the course development team in determining the optimal delivery methods. According to the literature, the documented challenges encountered during the pandemic encompassed a wide range of barriers, starting with overarching logistical issues such as border closures and limited internet access, and extended to individual student difficulties, including academic and digital literacy (Badrul et al. 2021).

When teaching online internationally, it has become clear that the digital divide has not been bridged and the global pandemic played a significant part in exacerbating it (Ryan and Nanda 2022). This exacerbation includes both larger-scaled issues such as the internet availability through to a more individual limited level of digital literacy. Internet availability obviously impacts participation in many aspects but also what resources can be used when connectivity is available, acknowledging bandwidth limitations. At the lower end of the scale is limited digital literacy which reduces effective engagement time as it is consumed with showing participants how to use new online tools and methods.

It is important to also point out that this difference also exists within nations in the form of rural–urban divide, a very real situation in PNG. Teachers participating in the course came from diverse regions across PNG (see Introduction to Part II). These ranged from the vast, thinly populated highlands with their mountainous terrain to the islands characterised by low population density. These regions exhibited varying levels of infrastructure development, particularly concerning internet access. Thus, when transitioning to a fully online mode of delivery, we had to take into consideration abovementioned factors, together with cultural factors. The course delivery also evolved during the pandemic to ensure continuity, responding to the teachers' individual circumstances by anticipating the impact of remote locations, lack of resources, absence of technology and poor connectivity.

At the broadest level, the course delivery modes in this project underwent four significant transitions due to border closures and subsequent travel restrictions, including the introduction of vaccination requirements. These shifts in modes of delivery highlight the agility and adaptability required from both teachers and academics to navigate the ever-changing circumstances during the pandemic. Furthermore, it required significant agility from AAPNG as well as intensification of necessary resources. It was this three-way partnership that helped to ensure the course's successful delivery.

This chapter, following on from the previous chapter, is also concerned with research. Chap. 3 addressed the first research question, that is, the more expected investigation of our course design and content. This chapter, less expected, addresses the impact of course delivery embedded in the second research question:

- *How did the modes of course delivery impact on teachers' learning?*

4.1 Course Delivery Modes in Graduate Certificate in STEM Education

Across the three cohorts, four delivery modes were implemented which we will refer to as Modes A, B, C, and D. This arrangement, and how it came about, was introduced in the Introduction to Part I.

The application of each mode varied between the cohorts. The enactment of each mode was dependent on factors such as international travel regulations, school holidays, PNG government elections, and project budget constraints. Each of the four modes involved varied levels of online and blended learning (Table 4.1). In Modes A, C, and D, the subjects were delivered in an intensive mode lasting 1 to 1.5 weeks. In all modes of the course, assessments were due on the last day of the teaching block for that subject (see Table 4.1).

4.1.1 Mode A: Aggregated on Campus (Intensive)

The face-to-face/blended mode delivered at our campus in Brisbane initially seemed to bypass certain technological obstacles due to the existing infrastructure, including internet connectivity, laptops, software, and technical support. Additionally, students had access to resources such as learning and library support and proximity to academic staff, which facilitated the development of academic literacies. Each day was divided into presentations (including guest presenters), workshops, group work, and independent study sessions. This mode also facilitated visits to schools that could highlight the application on integrated STEM approaches. Visits to centres that could show how STEM was applied in the real world were possible when the teachers were on our campus. For example, a visit to Substation 33 (https://www.substation33.com.au), a Brisbane-based social enterprise focussed on recycling digital waste gave teachers an opportunity to rethink how the UN agenda on sustainability could be embraced through STEM (UN, n.d.). Several teachers opted to pursue the project of Power Wells (https://www.powerwells.org/) in their schools (see Narratives 7.3 & 8.3). PowerWells is a project developed at Substation 33 to repurpose lithium-ion batteries to convert and store solar energy for use to power electrical devices.

4.1.2 Mode B: Distributed Online (Short Block)

Changes prompted by the global pandemic meant transitioning to remote online learning with individuals distributed across the nation of PNG. This mode was implemented as soon as the international and domestic borders were closed in March 2020, and the teachers had to work from their homes. Thus, our primary objective in this instance was to deliver the course online so that the teachers could engage from their home locations. The teachers had limited access to hardware. While some had a digital device (laptop or mobile phone), they were either too old or had technological issues (e.g., broken screens). The remote and rural locations of the teachers presented connectivity challenges. To overcome these barriers, the university team and AAPNG was able to persuade the Australian Department of Foreign Affairs and Trade (DFAT) to provide each teacher with a new laptop so that online learning could be implemented. The teachers and the university team were incredibly grateful for this support which was a first for such projects.

However, computers for remote online learning become less effective without internet connectivity. To mitigate this issue, the AAPNG took steps to provide support to teachers by offering high-speed dongles and mobile data allowances. This helped to enhance internet connectivity for those facing connectivity challenges. Overall, the provision of this hardware was a small investment (5%) compared to the project's overall cost Additionally, in cases where connectivity remained an issue, we took the initiative to ship learning resources, such as lecture slides, readings, and videos on USBs, as well as providing printed resources of the lecture slides and readings. This approach ensured that teachers had access to essential materials even in areas with limited or unreliable internet connectivity. These measures were implemented to facilitate a smoother learning experience and to support the teachers in their professional learning despite the obstacles posed by connectivity issues.

Table 4.1 Summary of the four modes used to deliver the course across three cohorts

Mode	Type	Study site	Delivery	Subject Duration (Weeks)	Instructor presence
A	Aggregated on campus (Intensive)	Australian University	Face-to-face & Blended Learning	1.5	Physical
B	Distributed online (Short block)	Teachers studying from their home locations in PNG	Online	6	Virtual
C	Aggregated online (Intensive)	Hotel in Port Moresby	Online	1	Virtual
D	Aggregated off campus (Intensive)	Hotel in Port Moresby	Face-to-face & Blended Learning	1.5	Physical

For most of the PNG teachers, digital literacy skills were basic and online learning was a new experience. To address this issue, we organised several workshops to improve the teachers' digital literacy skills. Each subject that was delivered in Mode B was allocated 6 weeks. During this period, teachers were working full time, either at their schools or from their homes depending on what the COVID-19 restrictions were in place at the time. Through PNG's National Department of Education, we were able to negotiate a study day (Wednesday) for the teachers with some of school principals when the course was running. On the study day, we scheduled a Zoom meeting (1.5–2 h) to discuss the key ideas from the learning modules and address any issues.

Considering remoteness, power surges, and connectivity issues, many teachers had to devise ways to overcome insurmountable problems to join classes. For example, one teacher climbed a mountain in Simbu Province in order to connect to a Zoom session (Fig. 4.1). Despite these barriers, attendance at the Zoom sessions was consistently high and in most instances was between 70 to 80%.

Fig. 4.1 Connecting to a Zoom session on a mountain side (Gumine, Simbu Province)

4.1.3 Mode C: Aggregated Online (Intensive)

Once the domestic borders in PNG re-opened, we switched to Mode C. It was also the time when international borders were still closed. Like Mode B, this was also delivered online but was delivered in an intensive mode. All teachers travelled to Port Moresby and the course was delivered in a comparable way to a traditional face-to-face course. A stark difference was that the university team was in Brisbane while the teachers were in a hotel in Port Moresby, just over 2000 kms away. AAPNG staff assisted with the coordination and some technical support was provided by some local technicians and the hotel staff. However, this intensive format was different to Modes A and B because the subjects had to be delivered in one week compared to one and a half weeks [Mode A] and six weeks [Mode B]. The risk of infection due to the COVID virus was one of the reasons for the shorter duration of this mode.

The teachers had to develop new knowledge and submit their assessments within this fleeting period. This online mode of delivery introduced different complications, including the implementation of various health restrictions within PNG, such as mask-wearing and room occupancy limits, which had an impact on both the delivery and learning experiences.

4.1.4 Mode D: Aggregated off Campus (Intensive)

When the international borders re-opened, the teaching team were able to travel to PNG and all teaching was conducted in a blended mode at a hotel in Port Moresby. In this mode, the teachers had the advantage of having access to the university team who provided guidance and support more effectively. The course delivery was intensive. However, unlike Mode C, the duration of the course was one and a half weeks. This change was directly due to the research feedback provided by the teachers about Mode C. Some felt that the five days (one week) was too short for the delivery of the subject. Hence, AAPNG supported the idea of extending the delivery time by 50% for each subject.

Through diverse approaches, it was evident that, considering the difficulties (e.g., border closures and risk of infections), a pragmatic combination of delivery modes ranging from aggregated to distributed, teaching durations ranging from intensive to short blocks, and instructor presence ranging from physical to virtual was the optimal balance for the learning environment. As the three cohorts progressed, the differing modes were implemented in varying proportions largely driven by the prevailing conditions (Table 4.2).

4.2 Impact of Different Modes on Teachers

In terms of teachers' overall performance in the course, there was no difference between the modes. The percentage of those who successfully completed their courses across the three cohorts was 98% despite variations in how the course was delivered. However, ongoing feedback from the teachers and the data gathered through our research showed that there was a difference in terms of the usefulness of each mode on their learning experiences. The impact of each of these modes on the participants, as revealed through our research, is discussed in the following section.

4.2.1 Mode A: Aggregated on Campus (Intensive)

In Mode A, the first and the third cohorts of participants engaged in face-to-face and blended delivery at our university (see Table 4.2). The times that the cohorts spent on campus were either 12.5% (Cohort 1) or 50% (Cohort 2). This mode of delivery provided teachers with hands-on experience and opportunities for immediate feedback from the teaching team. Additionally, the environment was ideal for hands-on activities and collaboration, which is especially important in STEM fields where experiments and demonstrations are common.

In-class support was an advantage of Mode A because it enabled the teachers to directly *interact with the lecturer* [Teacher 58]. Another explained that *at the end of the day, I prefer face-to-face learning. I am someone who asks a lot of questions and wants answers immediately to things I am unsure about* [Teacher 35]. In interview, William noted that the *lecturers were incredibly supportive and helpful.* Whenever *we needed some assistance* or had *some problems … with assessment or … with the way we handled things the lecturers were helpful.* William pointed out that *the sessions were very interactive,* and *our lecturers* kept us *updated on our progress by giving us feedback.* Miriam (interview) said that she was *more focussed on the course itself* with Teacher 23 offering that this focus was because of *no distractions from our family and our teaching duties.* For all teachers, returning to university was giant leap given that many had not engaged in any formal training after completing their undergraduate qualifications in PNG.

Table 4.2 Modes of delivery embedded in each cohort

Cohort	Percentage of subjects delivered through each mode			
	Mode A: Aggregated on campus (Intensive)	Mode B: Distributed online across PNG	Mode C: Aggregated online (Intensive)	Mode D: Aggregated off campus (Intensive)
1	12.5%	37.5%	50%	–
2		50%	50%	–
3	50%	–	–	50%

The learning environment in an Australian university was quite different to what the teachers had experienced as students. Despite this challenge, William (interview) believed that studying *in a new environment really motivated [them to push] beyond the limits* and this *mode was very good for us.* Michael (interview) said that studying in Australia also *took me away from family and ... work commitments, so ... I put in 100% effort into studying.* Consequently, some teachers believed that the on-campus mode *allowed* them *to complete tasks as I had ample time* [Teacher 31]. For William (interview), this mode gave him *a little bit exposure and ... the opportunity to do the things that I have never seen in my life as a teacher in the classroom. So, I believe that the face-to-face mode is the most convenient way for this course.* Lucy (interview) pointed out that she felt safer in Brisbane because safety was a concern for her when she attended the course in Port Moresby. She said that *in Brisbane there were no criminal things—like what we experience back home.* Safety was also a concern whenever she travelled from her province to Port Moresby. She also shared that in her province she was used to a poor internet connection which impacted on her capacity to engage with the tasks at hand. However, studying in Brisbane gave her access to an exceptionally reliable network which enabled her to easily engage with the online learning resources and complete her assignments.

One of the challenges for the teachers was that while their schools had released them to attend the course, their classes were still timetabled and running. No relief teachers were assigned to cover them while they were away. The teachers thus had to prepare work and students were expected to work independently with minimum supervision. This added pressure and created a distraction from their studies.

4.2.2 Mode B: Distributed Online (Short Block)

In Mode B, teachers engaged in online studying from their individual residences and workplaces across PNG. Cohorts 1 and 2 completed 37.5% and 50% of their course respectively through this mode of study. The transition to online learning from the comfort of their homes presented both opportunities and challenges. Some of the teachers appreciated the fact that they were given sufficient time (six weeks) to successfully complete each subject. More importantly, it gave them time to reflect on what they had learnt. The *online mode from school* [Mode B] *was better since I had a lot of time to go through the content carefully.* This teacher was also able to relate the new knowledge to his classroom because he *had the students around* him. In the minds of the teachers, having the students around enabled them to apply the ideas learnt in the course *immediately.* One of the teachers noted that online learning delivered *practical results* that were generated through the application of the latest ideas. This also *help*[ed] *me to do my assignment.* Another, Teacher 46 echoed similar views offering that *I applied the knowledge in my lessons at the same time* and *the impact of this is shown in my assessment results.* For this teacher, *it was it was fun learning new things, and then delivering it to the students 'right.'*

Another benefit of distributed online course delivery noted by the teachers was flexibility (e.g., Rice and Carter 2016; Veletsianos and Houlden

2019). In this case, the school and home were both familiar environments for the teachers. Despite working full time, studying, and attending to other matters, *learning from school/home was relaxing* [Teacher 44] because it enabled them *to attend to work and family matters* [Teacher 50] simultaneously. It was *more relaxing*, and it gave them more time *to do readings and workshop activities which gave us more understanding of the content* [Teacher 14]. Teacher 31 summed up how the distributed mode impacted on their learning:

> …. [it] *made me spend quality time in reading literature and understanding the presentations from a broader perspective. I also had printed materials which enabled me to read offline and jot notes and refer to them in the future. The subject was spread out over six weeks and offered me the flexibility to work to my satisfaction.*

Teachers also believed that online learning made them *really think* about the *individual work that I needed to do* (Elizabeth, interview). It immersed teachers in an environment where they became self-directed learners (Song and Hill 2007). This experience *helped me to be more self-directing, …. and reflect on what I'm doing,* (Elizabeth interview). Self-directed learning is a critical attribute of lifelong learning (Alexander et al. 2004) and is especially important in online learning environments (Carter et al. 2020).

Due to the persistent connectivity issues in some parts of the country, some of the teachers expressed their appreciation for the additional and alternative measures that were taken to ensure that they had access to learning resources and were not left behind. When making this provision, we believed that in the event of power or connectivity problems, learning resources on USBs would be a viable solution that would allow the teachers to continue their studies without interruption. The teachers found this effective because they *had printed materials for the subjects,* and they were *able to read in depth and understood things* (Michael, interview). Another teacher claimed to have *learnt a lot* online, but challenges such as *network issues, time constraints, juggling work and school* impacted her grades, pointing out that if it had *been face-to-face throughout, I would have learnt even more or even score better grades* [Teacher 33]. William (interview) believed that this approach to study was rife with *plenty of disturbances. There were kids and family at home and our family obligations took much of our time.* Consequently, *concentration and study times were very much forfeited. As a result, we had assignments done in the last minute and submitted.* As evident in these instances, it was not the course per se, but extrinsic factors that impacted on the teachers' successful participation in the course.

Managing online learning with a full teaching load was stressful for some teachers because the teachers had *a lot to complete* [Teacher 20]. Teaching and studying became a juggling act. For one teacher, *my students' work and needs were more* [pressing] *than my online study* [Teacher 54]. For most teachers, there was no change in their workloads. Thus, they had to meet their teaching obligations and were under a lot of pressure because most were teaching classes where their students were preparing for external examinations. In addition to the workload, the *lack of support from school administration* [Teacher 43] presented another challenge. A number were hard pressed for study time and consequently, *I honestly did not read and watch videos before the weekly Zoom meetings* [Teacher 54]. As one teacher noted, *there were times [when] I needed to fully read and understand the online resources* to *know more about what was needed to learn from each subject to improve my knowledge* [Teacher 54]. Thus, the combination of working full-time had a significant impact on the time that was available for studying.

4.2.3 Mode C: Aggregated Online (Intensive)

Cohorts 1 and 2 completed 50% of their course through aggregated intensive online mode which was delivered in two two-week blocks. In this mode, the teachers were based in a hotel in Port Moresby for two weeks at a time to complete two subjects. The university team delivered the course online from Brisbane. For most of the time, the technology support on the ground in Port Moresby was particularly good. While the hotel environment was extremely comfortable, it also took the teachers away from home and school and gave them a block of time for focussed study. The online setting from Port Moresby created challenges for many teachers although some teachers preferred the hotel environment. For example, one teacher said that *I was able to concentrate more in my hotel room than from home. I could also relate with my peers and the lecturers more* [Teacher 32]. Another teacher pointed out that *learning online while in Port Moresby generated less stress as I was away from active duty, so I had all the time to complete my course well* [Teacher 20]. As the *workshops brought* them *together* (Elizabeth, interview) there were also more opportunities for collaboration and communication because *I had colleagues with whom I shared and discussed ideas that helped make progress* and this could not be done when learning was conducted *in isolation* [Teacher 31]. The teachers appreciated *collegiality as* they *helped each other,* and this mode *encourage*[d] *good collaboration and group discussion between students, teachers and peers* (Gabriel, interview). The university team were online. However, in comparison with Mode B, there were a lot more opportunities for consultation to clarify *areas of doubt* [Teacher 31].

However, Mode C presented its own set of challenges. It was a *new style of learning which I believe was equally effective; however, one week was honestly too short (but we did fine regardless)* [Teacher 56]. This *intensive short turnaround time* and *due dates to the assignment* were a downer because it caused *pressure and fatigue* [Teacher 31]. Some teachers felt *exhausted* [Teacher 49] and *challenged* [Teacher 11] because of *time compression* and *a lot was covered in a short time* [Teacher 49]. One teacher drew an interesting analogy stating that the intensive course delivery was like *swallowing a thick book in one day* [Teacher 59]. One teacher stated that the study done in Port Moresby was *too intensive and was not helpful in the sense that due to time limitations, we were not able to do readings and workshop activities to grasp knowledge and skills but rather were under extreme pressures of getting the assessment tasks done* [Teacher 14]. Elizabeth (interview) *couldn't work around people* and found hard to *concentrate* as *we're in a room full of people who are not on the same wavelength, and who are moving around doing different things and talking.* Some teachers felt the intensive nature of the mode, impacted negatively on the quality of learning. One teacher shared her thoughts as follows; *honestly speaking, for me, less learning took place in the subjects that were covered in Port Moresby* [Teacher 14]. The consensus appears to be that Mode C was not well received. Without the physical presence of the university team, it lacked the collegiality and encouragement offered by the face-to-face experience of Mode A.

4.2.4 Mode D: Aggregated off Campus (Intensive)

Cohort 3 completed 50% of their course through this approach. This component was delivered in two three-week blocks. In many respects, it was like Mode C except that it was delivered over 1.5 weeks (instead of 1 week) and our university team travelled to Port Moresby to deliver the course (instead of online delivery). This additional time (7–8 days per subject with a weekend in between) and the presence of the university team in a face-to-face environment seemed to have addressed the issue of an *intensive short turnaround time* that was raised by teachers who went through Mode C. For Lucy (interview), Port Moresby was an *environment that was familiar to* her, and she was closer to her family. She believed that the off-campus mode did not impact on the quality of teaching. It was *the same* and in comparison, to the on-campus mode, they had similar access to the teaching team *in the morning till the afternoon* and were available *when we had questions.* Like most teachers, Lucy (interview) had to set work for her class during her absence. She was also expected to catch up and review the work with her students when she returned to school. She believed that this was much easier to do with the three-week block than when she went away to Australia for 6 weeks [Mode A].

4.3 Summary of Course Delivery and Impact

Teachers in the three cohorts experienced four different modes of delivery. All four modes of delivery shared a common objective, which was to enhance teachers' PCK and their understanding of an integrated approach to STEM in their contexts. Traditionally, as noted, Australia awards short courses are delivered in a face-to-face/blended mode on an Australian university campus [as in Mode A]. The COVID pandemic caused significant disruptions which were not only inconvenient but also posed numerous challenges and problems. However, the barriers enabled us to reimagine the course and think creatively. It enabled us to develop and implement Modes B, C, and D which were possibility firsts in terms of how such courses are delivered. We did not notice any significant differences in students' performances across the four modes. Feedback from the teachers demonstrated that all modes had their strengths and weaknesses. There were three key variables that influenced teachers' preferences—mode of delivery, study site, and duration. All other factors centred around these key variables.

4.3.1 Mode of Delivery

For the teachers in Cohorts 1 and 2, the two modes of delivery—face-to-face/blended Vs Online played a significant role. Some teachers viewed the shift to online learning as a "game changer". As one teacher wrote, *at first it was difficult,* but over time *I mastered the skills* [Teacher 16]. Overwhelmingly the responses showed that online learning enabled them to develop *critical time managemen*t [skills] *for task completion* [Teacher 51] and made them *read literature well before class and attempt related work* [Teacher 53]. They had to create a routine where they had *to do coursework apart from professional and private duties* [Teacher 31]. The course structure also made them *realise the need to work on assignment tasks immediately since the due dates were short and fixed* [Teacher 31]. The new *knowledge and experiences* [Teacher 61] impacted positively on their *digital literacy* because there were *a lot of new technologies like Zoom, and I was able to learn more from it* [Teacher 2]. Some teachers pointed out that there were *some technology issues* [Teacher 20] at times. This was compounded by the fact that there was no one close who was present *physically…to help* [Teacher 2]. But *this was also good because I learned a lot myself* [Teacher 2] and it *challenged me to give my best* [Teacher 15].

One teacher believed that using technologies for *online learning is a way forward* (Joseph, interview). It encouraged him *to change the approach to how* he *was teaching.* He got his *students to create a Google classroom and* he *was able to communicate with them during the pandemic,* thus studying online *helped* him *to think outside of the box …* and *I enjoyed it.*

Teacher 53 explained his preferences of the different modes of delivery as follows:

> As a learner, I think I did not face any big issues in adapting to new learning/delivery modes throughout the course because our QUT coordinating team were extremely helpful in ensuring easy transitions from one mode to another. I am a fast learner; hence I was able to grasp the skills and knowledge delivered throughout the course. Personally. I prefer the face-to-face mode and the remote online mode over six weeks because I had opportunity to ask direct questions and ample time to read and go through the recommended literature. I did not give my best in the online mode in Port Moresby as it was more intensive, and I feel that I did not fully understand the concepts delivered.

Teachers learned to *adapt* and use the different applications for which *I am grateful…and I will treasure it always* [Teacher 51]. One teacher summarised the importance of good planning when designing subjects for online delivery. Thus, *online learning was consistent in which every subject was smoothly covered as planned, and that kept me on the track* [Teacher 45].

Technical issues, such as unreliable network connectivity, were ongoing—not only in the teachers' home locations but also in the Port Moresby Hotel. *The biggest problem I have is the internet connectivity to do my studies* [Teacher 49]. Another teacher *really struggled to catch up with the rest of the participants and meet the assessment deadlines. Most of the time, the internet network was unstable. It is quite a challenge learning remotely* [Teacher 4]. Although at the same time the teacher acknowledged that the online *learning modes have made me aware of the different technologies that can be used to disseminate teaching and learning remotely, I have learned a lot of new things apart from the course itself* [Teacher 4]. For some teachers, the learning curve with digital technologies was quite steep. One teacher considered herself *computer illiterate* at the *start* of the course [Teacher 3]. This impacted on her capacity to keep pace with the course.

The feedback gathered from the teachers in this project demonstrates that there is no one size fits all in terms of modes of delivery. One admitted that *both face-to-face and online did satisfy the need that I had for learning about STEM Education* [Teacher 45]. Teacher 62 similarly did not like either approach because *the quality is not there. We were caught up with our normal teaching schedules and family issues in our schools. In hotels in Port Moresby, we were under a lot of pressure. Not enough … breathing space[time] to digest concepts.*

On the basis of the feedback provided by the PNG teachers, teacher upskilling should be underpinned by a mix of face-to-face, blended, and online approaches. They should have a chance to engage both synchronously and asynchronously. According to Kemp and Grieve (2014), "online and face-to-face activities can lead to similar levels of academic performance" (p.1). In this project, we also noticed that the modalities did not impact on teachers' academic outcomes. Each approach has benefits and constraints.

The teachers in this course have demonstrated that despite challenges (e.g., digital literacy issues), the teachers learnt to adapt. Delivering teacher professional learning in a face-to-face mode is not always feasible and can be costly. Taking them offsite and from their classrooms for short periods leads to other questions such as *Who will cover their classes when they are away? Who looks after family members in their absence*?

Upskilling teachers online has many benefits. Teachers can develop their knowledge and skills at a place and time of their choosing. It can also work in courses that involve hands-on engagement—such as our Graduate Certificate Course in STEM Education. However, for teachers to fully immerse in their learning, they need access to technologies that enable them to do their work and connects them to the internet. The provision of a laptop, a high-speed Wi-Fi data access by AAPNG and DFAT resolved many issues for teachers. It optimised teachers' participation in online learning and created pathways to ensure that no teacher was left behind. The real variable in teacher preference between the different modes of delivery was that teachers' classes were not covered in their absence. This made their full engagement in any of the modes close to impossible.

4.3.2 Study Sites

There were three study sites—university campus, hotel, and home/school. The University campus (in Australia) was noted for the convenience of having our university team and other resources on site. This enabled the teachers to get support and responses to the queries and questions as needed. They also had a block of time to focus on their studies and were free from family, school, and community-related disruptions. The teachers also got to experience a learning environment that was quite different to what they were used to back in their home country. While not explicitly stated in the data, it was evident that the teachers also enjoyed Australian culture and took opportunities to visit places and watch football matches. PNG citizens are very fond of Australian rugby league.

While studying on an Australian campus has many advantages, it can be costly. Expenses include travel, accommodation, insurance, and teachers' *per diem* allowance. To participate in the course, the teachers had to be away from their families. Anecdotal and research data suggested that when some of the teachers were away studying, their positions were not backfilled. This presented challenges to the teachers because they had to catch up on what the students had missed. This complicated the issue given that many were secondary school teachers and their students were preparing to sit for external exams.

Delivery from the hotel in Port Moresby gave the teachers a chance to get a block of time for their study. A distinct difference was that the study block at the hotel was shorter and was of three weeks' duration as opposed to the on-campus mode which was over six weeks. This was an advantage because the teachers were away from their schools (and homes) for shorter periods. Hence, catching up on any work that students missed was easier. Running the course in Australia versus running it in Port Moresby may have also impacted on the overall costs.

Studying online from home/school had its own basket of rewards and challenges. For some of the teachers, the comfort of studying from home was a bonus. They could study, teach their students, and attend to the needs of their families. The study mode was distributed which meant that their study was spread over a few weeks. Many believed that they could apply what they were learning more readily through this approach. However, some of the teachers found this difficult because there were too many interruptions, and they did not get uninterrupted periods to focus on their study. Nonetheless, the biggest challenge in this instance was connectivity. Not all teachers had reliable internet connections from the home and school locations which impacted on the quality of their learning.

Delivering an in-service course from different venues has merit. Traditionally, award courses are delivered from an Australian University. A mixed mode with parts delivered in Australia and offshore suits teachers and may be economically more viable. For such an approach to work, the course needs to be delivered in short blocks.

4.3.3 Duration of Delivery

The subjects in the course were delivered, either over an intensive (1 to 1.5 weeks) or over a longer period (6 weeks). Much of this was determined by Australia Awards in consultation with the PNG National Department of Education and the schools. The complexities associated the COVID-19 virus also played a significant part. On campus, each subject is usually delivered over 5 weeks with class contact time of 2 to 3 hours per week. This generally suits the students because they are full-time teachers and have other responsibilities. All learning resources are pre-loaded and designed for self-paced learning. One of the challenges is to determine how such a course can be delivered in an intensive mode. There is value in this approach because learners can complete the course requirements in a short time. However, determining the optimal time for delivery is difficult and can only be fine-tuned through learner feedback. Most of the PNG teachers found the 1-week intensive mode was challenging because there was too much to absorb and digest. The university team was based in Australia and this distance impacted on the level and immediacy of the support that the participants were receiving. However, the teachers in Cohort 3 felt that 1.5-week mode was more manageable. It had a weekend in between which gave them time to revise and reflect which effectively extended the learning cycle to 9–10 days.

Determining the optimal amount of time for course delivery is challenging. In-service teachers are adult learners with varied experiences and backgrounds. Thus, the university teams had to continually review academic performance and act on feedback. They also have to be alert to extrinsic factors impacting on the learners' progress.

It would seem that many teachers in this course preferred face-to-face delivery. The reasons for this preference tend to rest with the social aspect of learning in the company of peers and with immediate access to support and resources. Where there were difficulties with any of the modes, it appears to be extrinsic to the course itself. The main barriers were about continuing responsibilities at school and at home. Where there were technical issues, extraordinary solutions were found to provide individuals with the equipment and access they needed. By being placed in the unique position of experiencing four different modes of course delivery in the one course, we have gained valuable lessons in how best to deliver courses to PNG and other developing nations.

References

Alexander S, Kernohan G, McCullagh P (2004) Self-directed and lifelong learning. Stud Health Technol Inform 109:152–166

Allen IE, Seaman J (2013) Changing course: ten years of tracking online education in the United States. Babson Survey Research Group

Aung TN, Khaing SS (2016) Challenges of implementing e-learning in developing countries: a review. In: Zin T, Lin JW, Pan JS, Tin P, Yokota M (eds) Genetic and evolutionary computing. GEC 2015. Advances in Intelligent Systems and Computing, vol. 388. Springer. https://doi.org/10.1007/978-3-319-23207-2_4

Australian Government (n.d.) Australia Awards. https://www.dfat.gov.au/people-to-people/australia-awards

Badrul HK, Saida A, Soheil Hussein S, Zuheir Najee K (eds) (2021) Challenges and opportunities for the global implementation of e-learning frameworks. IGI Global. https://doi.org/10.4018/978-1-7998-7607-6

Carter RA, Rice M, Yang S, Jackson H (2020) Self-regulated learning in online learning environments: strategies for remote learning. Inf Learn Sci (5/6):321–329. https://doi.org/10.1108/ils-04-2020-0114

Castle SR, McGuire CJ (2010) An analysis of student self-assessment of online, blended and face-to-face learning environments: Implications for sustainable education delivery. Int Educ Stud 3(3):36–40. https://doi.org/10.5539/ies.v3n3p36

Chandra V (2019) Share engage educate—SEEding change for a better world. Brill

Creswell JH (2014) Educational research: planning, conducting, and evaluating quantitative and qualitative research, 4th ed. Pearson

Hodges C, Fowler D (2020) COVID-19 crisis and faculty members in higher education: from emergency remote teaching to better teaching through reflection. Int J Multidiscip Perspect High Educ 5(1):118–122. https://doi.org/10.32674/jimphe.v5i1.2507

Kemp N, Grieve R (2014) Face-to-face or face-to-screen? Undergraduates' opinions and test performance in classroom vs. online learning. Frontiers in Psychology, 5, 1278

OECD (Organisation for Economic Co-operation and Development) (2020a) Strengthening online learning when schools are closed: The role of families and teachers in supporting students during the COVID-19 crisis. OECD Publishing

OECD (Organisation for Economic Co-operation and Development) (2020b) Learning remotely when schools close: how well are students and schools prepared? Insights from PISA. OECD Publishing

Oko J (2021) An investigation into factors affecting Grade 10 and 12 students' Mathematics performance in Port Moresby, Papua New Guinea (PNG) (Doctoral dissertation). https://hdl.handle.net/2440/134190

Rice MF, Carter RA Jr (2016) Online teacher work to support self-regulation of learning in students with disabilities at a fully online state virtual school. Online Learning 20(4):118–135

Ryan M, Nanda S (2022) COVID-19: social inequalities and human possibilities. Routledge

Song L, Hill J (2007) A conceptual model for understanding self-directed learning in online environments. J Interact Online Learn 6(1):27–42

Veletsianos G, Houlden S (2019) An analysis of flexible learning and flexibility over the last 40 years of distance education. Distance Educ 40(4):454–468. https://doi.org/10.1080/01587919.2019.1681893

Wranga E, Aku G, Otima F, Chandra V (2021) Teaching science using digital technologies: some insights from Papua New Guinean schools. Teach Sci 67(3):21–24

Zhang B, Jin Z, Peng Z (2018) Bridging the digital divide: making the digital economy benefit to the entire society. IDEAS Working Paper Series from RePEc. https://ideas.repec.org/p/zbw/itsb18/190412.html

Senka Henderson is an Associate Professor at QUT, a Senior Fellow of the Higher Education Academy (HEA) and an Associate Fellow (Indigenous) of the HEA. She is the Course Co-ordinator of the Master of Teaching (Secondary) programme. Senka teaches multiple curriculum, assessment, and pedagogy units in secondary studies across both undergraduate and postgraduate courses. She was involved in the case study research which underpins Part I. She contributed to Chap. 3 and led the writing of Chap. 4. Along with Charisse Farr, she also wrote the Introduction to and Review of the chapter on *Authentic Learning*.

Vinesh Chandra is an Associate Professor and the STEM Subject Area Co-ordinator at Queensland University of Technology (QUT). In 2024, he was awarded an Australian Award for University Teaching (AAUT) for his contributions to advancing STEM education in Australia and overseas. He also served as one of the leaders for the *PNG STEM Education Teacher In-service Project*. In addition to being co-editor of this book, Associate Professor Chandra authored this chapter and contributed to others. He has also co-authored other books on STEM education, including the award-winning "STEM Education in the Primary School: A Teacher's Toolkit" published by Cambridge, and "My STEM Workbook" for primary school students.

Matthew Readette is a sessional academic, senior research assistant and a Fellow of the Higher Education Academy (HEA). He has recently completed his PhD and has taught various units and short courses focussed on teacher education frameworks for global education, as well as academic and digital literacies. His research focuses on social and spatial mobility, international education and teacher education. He was involved in the research underpinning Part A and has made contributions to Chaps. 3 and 4.

Dann Mallet is Professor of STEM Education and Professor of Mathematics at Queensland University of Technology. He is experienced in academic and curriculum team leadership and mentoring, as well as extensive experience as a university Mathematics and STEM educator at all levels. Dann has initiated and led significant international education aid projects in the Indo-Pacific region, and has been recognised through numerous awards and nominations, including, most recently, a Gold in the QS Wharton Awards and High Commendation in the IEAA awards. Aside from his role as co-editor, he has contributed to Chaps. 3 and 4. Along with Vinesh Chandra, he wrote the Introduction to and Review of Chap. 8 *Project-Based Learning*.

5 Implications for Professional Learning and Leadership

Vinesh Chandra

We have thus far described the philosophy of STEM Education and explained how our approach to the design of the Graduate Certificate in STEM Education was based on a four-phase cyclical learning experiential model (Kolb 1984; Kolb and Kolb 2009). We aimed to offer course participants the opportunity to learn through concrete experience, reflective observation, abstract conceptualisation, and active experimentation.

This part, through Chaps. 3 and 4, further reported on our case study research which looked at how the *design* and *delivery* of the course impacted on the teachers who participated. Analysis of the survey data and interview responses produced valuable insights. Additionally, we drew upon the data to determine several implications that should be considered in future practice.

V. Chandra (✉)
Faculty of Creative Industries, Education, and Social Justice, School of Education, Queensland University of Technology, Kelvin Grove, QLD, Australia
e-mail: v.chandra@qut.edu.au

5.1 In-Service Professional Learning Should Live Up to Teachers' Expectations

Teachers enrolled in the STEM Education course for two key reasons. First, they wanted to improve the quality of teaching in their classrooms by upgrading their pedagogical content knowledge (PCK) and skills. Second, they wanted to learn more about integrated STEM Education because of the PNG government's directive. Evidence from our research clearly showed that the course was fulfilling expectations that the teachers had when they initially signed up. William (interview) summed up his thoughts as follows:

> This course made me think differently. It challenged me to shift from a traditional system of teaching to taking a more authentic approach. Bringing teaching and learning outside of the classroom walls and making the real-world connections.

Thus, we concluded that teacher in-service courses should be clearly aligned with teachers' initial expectations when they sign up for such courses. The high levels of satisfaction reported in the survey would suggest that, in this case, teachers' expectations were met.

V. Chandra et al. (eds.), *Integrated STEM Education in Papua New Guinea*, Sustainable Development Goals Series, https://doi.org/10.1007/978-981-95-0828-0_5

5.2 In-Service Professional Learning Should Be Contextually Appropriate

This is particularly the case with programmes that are delivered as Official Development Assistance (ODA), usually by a team from a more developed and wealthy country to a country perceived to need aid. Quite often, it is assumed that what works in one country will also work in another. Sometimes, unfortunately, there is a failure to realise the value of the STEM knowledge and skills that exist within traditional and cultural practices in all communities throughout the world. There is an erroneous but abiding perception that only knowledge generated in the developed world is worth knowing. In classrooms, an integrated approach to STEM Education can level the playing field if the teachers are skilled with contextually appropriate pedagogical toolsets. One of the ways to achieve this is to ensure the course development team has or gains broad knowledge of the context. Elizabeth (interview) felt that the contextual connection was *done well* for the course discussed in this book. The course design took into consideration some of the facts such as *we were different. we're not as Westernised as Australia, we don't have very many industries ... we're all from different parts of Papua New Guinea and all different*. The course was adapted and adjusted to suit the teachers' needs. *I think the benefit of having [the program leader] coming from one of the Pacific Islands* added value because he knew *PNG [and] how we [learn]...* so *I think it was done very well, and... we're grateful for that.*

5.3 In-Service Professional Learning Courses Should Allow Participants to Practise Their Learning in an Ongoing Way

Designing the in-service STEM Education course to have learning occurring in blocks is an effective strategy because it provided opportunities for the teachers to practise their learning on a continuous basis. It created a three-way connection between teachers' concrete learning experiences in the course, concrete experiences of knowledge application in their classrooms, and their reflections on those experiences (as aligned to the cyclical experiential model). Teachers had the opportunity to reflect on their experiences both at a personal level and with their contemporaries enrolled in the course. This opportunity can also be incorporated into the design of the course (e.g., workshops) where the teachers share their contextual experiences.

5.4 In-Service Professional Learning Courses Should Engender Personal Growth and Self-Esteem

In-service courses should be such that they buoy teachers' confidence and elevate them to another level within their schools and communities. The in-service course in this project initially developed the teachers' pedagogical content knowledge (PCK) and progressively shifted their capacities to apply this knowledge within their contexts. The underlying rationale of the course was to not only upskill the participants as STEM teachers but also to support their leadership in STEM education. The final subject in the course, for instance, created an opportunity for the teachers to present their visions and action plans for an integrated STEM activity. All teachers completed this task successfully and the majority have implemented their plans. Joan (interview) shared her experiences as an emerging leader as follows: *What I gained from this short course...helped my school very much. I presented my knowledge to the teachers.* Joan believed that there were *many outstanding impacts.* Joan's school was *piloting the new Standards-Based Curriculum...and this STEM course* had a positive impact on the reform exercise *especially in programming teaching and assessment.* As a leader directing change, she was also noticing the difference *high technology integration and various teaching methods connected with classroom teaching* was making in her school.

5.5 Teacher Selection for In-Service Professional Learning Should Be Teacher Driven

The responses from the survey and interviews demonstrated that the teachers *wanted* to do the course. No one had been required or pressured to enrol. This willingness could be one of the drivers for success. Teachers who want to engage with in-service training are also more likely to apply their knowledge and skills in their classrooms following completion of the training. They therefore came with an innate purpose, an "autonomous motivation". Teacher 32 explained that: *I wanted to make a difference in the learning taking place in my classroom. Learning was not taking place as it should be, so when I saw this opportunity, I thought this would make a difference in my approach to teaching and learning for my students, so I enrolled.*

5.6 Course Designers Need to Be Aware of Teacher Responsibilities Outside of the Course

The PNG teachers enrolled in the Graduate Certificate in STEM Education remained responsible for their classes in PNG, even while they were away. For example, when in Brisbane, Lucy was *still on the teaching timetable, juggling study and work,* and was *sending work back for her students.* She felt that *she was invisibly present* in her school. Even though Lucy had set work for the students so they could work independently, six weeks of absence from her students took its toll. She was monitoring them and noted that she emailed *formative tests for them, but they all did badly.* Upon return to her school, she was expected to *catch up and go over lessons with the students* that they had been expected to cover during her absence and, at the same time, cover the new work that was at hand. Being away from home was also challenging for Lucy because she had a young child and her absence from *was putting pressure on the father* with the daily chores and responsibilities such as taking the child to school. All in all, she felt that it was *too long ... to be away from home.*

Part I has introduced how a course came together based on a goal to make positive change in a developing country and how it succeeded despite all manner of barriers put in its way. It has ended with a set of implications for practice in designing and conducting this type of in-service teacher professional learning. Interestingly, the implications would ring true in any part of the world for any cohort of teachers engaging in professional learning. Part II will share the positive outcomes of the course through a selected set of teacher narratives grouped by STEM pedagogies.

References

Kolb DA (1984) Experiential learning: experience as the source of learning and development, vol. 1. Prentice-Hall

Kolb AY, Kolb DA (2009) Experiential learning theory: a dynamic, holistic approach to management learning, education and development. In: Armstrong SJ, Fukami CV (eds) The Sage handbook of management learning, education and development (pp 42–68). Sage. https://doi.org/10.4135/9780857021038.n3

Vinesh Chandra is an Associate Professor and the STEM Subject Area Co-ordinator at Queensland University of Technology (QUT). In 2024, he was awarded an Australian Award for University Teaching (AAUT) for his contributions to advancing STEM education in Australia and overseas. He also served as one of the leaders for the *PNG STEM Education Teacher In-service Project.* For this project, the team was awarded a Gold in the QS Wharton Awards and a High Commendation in the IEAA awards. In addition to being co-editor of this book, Associate Professor Chandra authored this chapter and contributed to others. He has also co-authored other books on STEM education, including the award-winning "STEM Education in the Primary School: A Teacher's Toolkit" published by Cambridge, and "My STEM Workbook" for primary school students.

Part II
Praxis—Teacher Narratives

The previous section of this book (Part I) documented teachers' professional learning and described the experiences of the secondary school teachers from Papua New Guinea undertaking a Graduate Certificate in STEM Education in Australia (see Chapters 2–5). This part, Part II (Chapters 6–9), is about teacher praxis, that is, it is concerned with how STEM teaching practice has been impacted by professional learning.

In the final subject of the STEM Education course (see Table 1), the PNG teachers were asked to create an action plan of how they would integrate STEM into their schools. A selection of these action plans has been enacted and submitted as narratives for this book to show how the teachers responded to curricular demands, met the specific needs of their communities, overcame constraints, and reflected the pedagogical frameworks which support teaching and learning in STEM. In undertaking this task, they became teacher authors. Some are recounts of a completed activity while others are a taster indicating that an activity has begun but, for a range of reasons has not been seen through to completion. Most wrote about projects or experiments, that is, the first time they had attempted an activity or approach. Lessons were learned about what paths might be followed in future iterations of the experiments. In all, we share eighteen narratives from sixteen of the dedicated teachers who undertook the STEM Education course. They are grouped into four chapters with each representing a discrete contextualisation of STEM education dependent on priority and circumstance. Unique aspects emerged inductively from the narratives. They are: *Authentic Learning* (considered in Chapter 6); *Design Thinking* (Chapter 7); *Problem-Based Learning* (Chapter 8); and *Teacher Leadership* (Chapter 9).

Each chapter in this part, in turn, presents four or five teacher narratives. The narratives of each chapter are introduced with a description and definition of the chapter focus written by academics who were involved in teaching the graduate course in STEM Education. Each chapter is concluded with a review of the aspect, analysing, and synthesising the teacher narratives.

Each narrative presents as a praxis. Put simply, praxis is practice informed by theory. It has been defined as "the interdependence and integration – not separation—of theory and practice, research and development, thought and action" (Zuber-Skerritt, 2001, p.15). The term has been deliberately chosen

Table 1 Structure of narrative chapters.

'What is' *Academic Authors*	Teacher Narratives *PNG Teacher Authors*	Review *Academic Authors*
Research background to the aspect of STEM Education	Each narrative is described in terms of its: Rationale Implementation Outcomes Reflection	Analysis and synthesis of teacher narratives to profile the aspect in the context of PNG education.

here because of its emphasis on action or "doing". Mahon et al. (2020) explained that:

… the word *praxis* has been ascribed meanings which distinguish it from ordinary everyday understandings of practice. ... praxis as *action* [is] based on deliberation about its potential consequences, and oriented by commitments to contribute positively and meaningfully to society, and to the good of humankind. (p.18)

The concept of praxis also has relevance in this instance because of its inherent moral component. Kemmis and Smith (2008) stressed the purpose of praxis as:

… action that is morally committed, and oriented and informed by the traditions of the field. ... Praxis is what people do when they take into account all the circumstances and exigencies that confront them in the particular moment and then, taking the broadest view they can of what is best to do, they act. (p.4)

The teachers whose narratives are presented in this part of the book share a deep concern for their students and their futures. They are also concerned with the future of their country, their communities, and the sustainability of their lives. Each has taken the context of their schools, communities, and the lives of their students into account in the pedagogical and curricular decisions they make. Because praxis is "always particular", that is, it describes:

… action taken in, and in response to, concrete conditions of place, time, material arrangements, and particular circumstances in this sense, educational praxis is situated (i.e., context-dependent) and responsive. It is shaped by, and also influences/shapes, practices (e.g., teaching practices and learning practices) and arrangements (e.g., classroom set-ups, arrangements of bodies, student solidarities).

(Kemmis et al., 2020, p. 22)

Thus, the situated narratives in this book must be read in the context of life in contemporary Papua New Guinea. It is important to note, for instance, that, in 2019, PNG was ranked 155th out of 189 countries on the UN's Human Development Index (UNDP, 2020), a summary measure of average achievement in the following key dimensions:

…a long and healthy life, being knowledgeable and having a decent standard of living. The education dimension is measured by mean of years of schooling for adults aged 25 years and more and expected years of schooling for children of school entering age.

(UNDP, 2023, para. 3)

Like many developing countries, PNG has its own challenges that significantly impact on its education system. Some of the challenges include limited access to quality education, infrastructure deficiencies, and regional disparities. Like many developing countries, the quality of education in Papua New Guinea varies across regions. Urban areas have better-resourced schools and more qualified teachers, while rural and remote areas face shortages of teachers, textbooks, and basic facilities.

As will become apparent through the narratives, the current situation is in need of improvement. Not every school in PNG has access to clean water or sanitation facilities. One of the goals of the PNG Department of Education's *National Education Plan* was to provide 50% of schools with functional water facilities by 2017, expanding to 70% by 2019 (see PNGDOE, 2016). Only 13% of the population has access to electricity (PNG PM Marape, 2020; PNGDOE, 2016). Disparities such as these across the country impact the overall quality of education. The teacher authors in this book work in this context, and most have used their circumstances as the catalyst for hands-on learning experiences for their students.

The work of the teachers aligns with the achievement of the Sustainable Development Goals (SDGs) (UN, n.d.). The 17 goals (Fig. 1, Appendix A) summarise the UN's "shared vision of humanity and a social contract between the world's leaders and the people … to wipe out poverty, fight inequality and tackle climate change" (UN, 2015, para. 2). The interdisciplinary nature of STEM fosters the necessary innovative, critical, creative, and collaborative ways of thinking and problem-solving to help face some of the world's biggest challenges outlined in the SDGs. Partnering STEM Education with the SDGs as a framework provides the opportunity for deeply contextualised learning that targets solving these problems at local levels while contributing to global efforts. Appendix A provides a detailed account of how the teachers in this book have contributed to the achievement of various goals.

The narratives cover teaching and learning experiences in a wide range of age groups and discipline areas. While the focus is on integrated STEM activities, we thought it would also be useful to map how the narratives relate to curriculum content (see Appendix B).

In addition, we have added a summary table to each narrative which draws on the STEM-PBL framework proposed by Forbes et al. (2021). The framework has five steps: Project Identification; Curriculum Connections; Assessment Ideas; Teaching and Learning process; and, Reflection. Our tabular summaries in the first three chapters of Part II offer: Problem (analogous to Project Identification); Connections (equates to Curriculum Connections); Pedagogy (analogous to Teaching and Learning Process); and Reflection. The fourth chapter, on Teacher Leadership in STEM, replaces Pedagogy with Domain—a reference to York-Barr and Duke's (2004) categorisation of teacher leadership practices.

In sum, the narratives presented in this part of the book complete Kolb's (1984) experiential learning cycle by encouraging the teachers to document their active experimentation and reflection (see Chapters 1–2). Writing has provided them with the opportunity to communicate and to share their experiences. They became cast in the role of action researchers looking to improve

Fig. 1 Sustainable Development Goals (SDGs). (UN, n.d.). Image source: https://www.un.org/development/desa/pd/content/sustainable-development-goals-2

their own practice and influencing others in their schools. In fact, many of the teachers have found other ways to communicate, usually in the form of leading professional in-service training in their own schools and further afield. The narratives and their shared intention to continue their projects are clear instances of teachers being active participants in a cyclical experiential learning model (see Figure 1). The learning they have started, through the course in STEM Education and continued, often through their own endeavours, has not been closed or static. They have transferred their learning to their practice and to the practice of their colleagues.

The teacher narratives are presented in four chapters, each explicating a discrete aspect of STEM Education and showcasing the richness of applications that the teachers have brought to their classrooms in PNG. The first chapter focuses on Authentic Learning (Chapter 6), the second on Design Thinking (Chapter 7), the third looks at Problem-Based Learning (Chapter 8), while the final chapter discusses Teacher Leadership (Chapter 9). While the chosen pedagogies have a distinct character, the first three share a focus on action or hands-on learning. There are similarly nuanced connections between these and such published pedagogies as: Action Learning, Problem-Based Learning, Experiential Learning and Inquiry Learning (see Albion, 2015; Kirschner et al., 2006).

Together the chapters can be seen to represent Kolb's abstract conceptualisation of STEM Education and how it can be applied to any learning environment. The narratives present a range of concrete experiences to aid student learning in STEM and demonstrate how teachers have merged their prior knowledge (of content and curricular requirements) into unfamiliar pedagogies. The narratives are not exclusive or intrinsic to Papua New Guinea. Rather, they are teaching experiments and learning experiences here which have transferable messages to schools, teachers, and trainers of teachers everywhere.

section

ReferencesAlbion, P. (2015). Project-, problem-, and inquiry-based learning. In M. Henderson & G. Romeo (Eds.), *Teaching and digital technologies: Big issues and critical questions* (pp. 240–252). Cambridge University Press

Kemmis S., Edwards-Groves, C., Jakhelln, R., Choy, S., Wärvik, G-B., Gyllander Torkildsen, L., & Arkenback-Sundström, C. (2020). *Teaching as Pedagogical Praxis*. Springer

Kemmis, S., & Smith, T. (2008b). Praxis and praxis development. In S. Kemmis & T. Smith (Eds.), *Enabling praxis: Challenges for education* (pp. 3–13). Sense

Kirschner, P. A., Sweller, J., & Clark, R. E. (2006). Why minimal guidance during instruction does not work: An analysis of the failure of constructivist, discovery, problem-based, experiential, and inquiry-based teaching. *Educational Psychologist, 41*(2), 75–86. https://doi.org/10.1207/s15326985ep4102_1

Kolb, D. A. (1984). *Experiential learning: Experience as the source of learning and development* (Vol. 1). Prentice-Hall

Mahon, K., Heikkinen, H.L., Huttunen, R., Boyle, T., Sjølie, E. (2020). What is educational praxis? In K. Mahon, C. Edwards-Groves, S. Francisco, M. Kaukko, S. Kemmis, & K. Petrie (Eds.), *Pedagogy, Education, and Praxis in Critical Times* (pp. 15–37). Springer Singapore. https://doi-org.ezproxy.library.uq.edu.au/10.1007/978-981-15-6926-5_2

PNGDOE (Papua New Guinea Department of Education). (2016). *National Education Plan 2020-2029: Quality learning for all.* https://www.education.gov.pg/admin/images/products/b98c46340f30348570b694bc4af6eae5.pdf

PNG PM Marape launches Schools of Excellence Policy (2020, June 15). Papua New Guinea Today. https://news.pngfacts.com/2020/06/png-pm-marape-launches-schools-of.html

United Nations (UN). (n.d.). *Sustainable development goals.* https://www.un.org/sustainabledevelopment/sustainable-development-goals

United Nations (UN). (2015). *Launch of new sustainable development agenda to guide development actions for the next 15 years.* http://un.org/sustainabledevelopment/blog/2015/12/launch-of-new-sustainable-development-agenda-to-guide-development-actions-for-the-next-15-years

UNDP (United Nations Development Programme). (2020). *Human Development Report 2020.* Retrieved https://hdr.undp.org/sites/default/files/Country-Profiles/PNG.pdf

UNDP (United Nations Development Programme). (2023). *Explore HDI (Human Development Index).* https://hdr.undp.org/data-center/human-development-index#/indicies/HDI

York-Barr, J., & Duke, K. (2004). What do we know about teacher leadership? Findings from two decades of scholarship. *Review of Educational Research, 74*(3), 255–316. https://doi.org/10.3102/00346543074003255

Zuber-Skerritt, O. (2001). Action learning and action research: Paradigm, praxis and programs. In S. Sankara, B. Dick, & R. Passfield, R. (Eds.), *Effective Change Management through Action Research and Action Learning: Concepts, Perspectives, Processes and Applications* (pp. 1—20). Southern Cross University Press

Authentic Learning in STEM

6

Senka Henderson, Charisse Farr, Paula Niandros Polume, Debbie Zaccias, Gideon Aku, Anastasia Paraka, and Koni Wamzil N'Drihin

Abstract

This chapter presents the narratives of five teachers in PNG who implemented authentic learning through integrated STEM projects. Despite facing significant challenges, such as limited resources, overcrowded classrooms, and unreliable infrastructure, these teachers designed innovative, community-driven learning experiences. Their projects addressed pressing local issues, including poverty, health, and waste management, allowing students to apply their knowledge in meaningful, real-world contexts. The narratives highlight how creativity, resilience, and local resourcefulness can overcome systemic barriers, empowering students with practical skills and a strong sense of agency. By aligning education with students' lived realities, these teachers demonstrate that authentic, transformative learning is not only achievable but essential in under-resourced settings, offering valuable insights for global education in similar contexts.

Keywords

Authentic learning · STEM integration · Global issues · SDGs

S. Henderson (✉) · C. Farr
School of Education, Queensland University of Technology, Kelvin Grove, QLD, Australia
e-mail: s10.henderson@qut.edu.au

P. N. Polume
ECOM Secondary School, Lorengau, Papua New Guinea

D. Zaccias
Bumayong Lutheran Secondary School, Lae, Papua New Guinea

G. Aku
Utmei Secondary School, Kokopo, Papua New Guinea

A. Paraka
Park Secondary School, Mount Hagen, Papua New Guinea

K. W. N'Drihin
National School of Excellence, Lae, Papua New Guinea

6.1 What Is Authentic Learning in STEM?

Senka Henderson and Charisse Farr

Authentic learning (AL) is a teaching method that situates learning within real-world situations (Herrington et al. 2014). In AL, students engage in the same problem-solving challenges in the curriculum as they would encounter in their daily lives, linking classroom learning to real-life problems and situations that are meaningful to them (Lombardi 2007; Reeves et al. 2002). Studies have shown that students prefer action over passive listening and most educators consider learning through doing to be the most effective method

V. Chandra et al. (eds.), *Integrated STEM Education in Papua New Guinea*, Sustainable Development Goals Series, https://doi.org/10.1007/978-981-95-0828-0_6

(Moeed and Anderson 2018). However, only recently have authentic activities gained significant attention, driven by the growing emphasis on STEM disciplines in schools, and the development of modern technology and engineering.

There has been considerable research conducted recently on authentic learning and authenticity in education. However, despite the demonstrated benefits of authentic student activities, opportunities for their implementation in schools are often limited. This is largely due to curricula being overcrowded with content, leaving insufficient time for open authentic activities. Additionally, many hands-on activities conducted in schools may not be considered authentic activities. There is a need for schools to consider ways to incorporate more authentic student activities into their curriculum to enhance student engagement, motivation, and learning outcomes (Herrington and Parker 2013; Reeves et al. 2002).

The concept of authenticity is ubiquitous and often contested (Anker-Hansen and Andreé 2019). The notion of authentic learning in education is also used loosely and inconsistently in literature (Nachtigall et al. 2022; Palm 2008). In this chapter, we adhere to a socio-cultural conception of authenticity, defined as active student participation in practices and activities that have a meaningful context and relevance to students. We use the word "authentic" in association with student learning in general, with activity, classroom environments and in relation to the nature of the 'real' world beyond the classroom and in relation to student personal meaning making.

To design authentic learning activities, teachers use different pedagogical approaches such as inquiry-based, project-based, service-based, and problem-based learning. These encourage students to engage in open inquiry that simulates real-world scenarios (Forbes et al. 2021). Students are given the opportunity to explore, experiment, analyse, evaluate, and synthesise information to solve problems, often in collaboration with their peers. As these tasks are connected to real-life problems and situations, students are developing essential 21st-Century skills that prepare them for future employment, such as critical thinking, creativity, communication, collaboration, and teamwork (Reeves et al. 2002). Moreover, engaging in authentic activities promotes deeper understanding and long-term retention of concepts, as students take ownership and become more actively involved in their own learning (Moeed and Anderson 2018).

Authentic activities also play a significant role in promoting the Sustainable Development Goals (SDGs), as students are encouraged to explore real-world problems and issues related to ending poverty, improving health and education, reducing inequality, and tackling climate change while preserving natural resources (UN, n.d.). Through SDG-related activities, students can learn to identify global problems and find solutions for local issues. These activities also help students become active and engaged citizens who are committed to creating a more sustainable and equitable world (Chopra and Banerjee 2022). Examples of authentic learning are presented in this section through five teacher narratives:

TEACHER NARRATIVE 6.1: Paula Niandros Polume led her Grade 9 Mathematics students at ECOM (Evangelical Church of Manus) Secondary School at Lorengau in Manus Province through the making of pencil cases from recycled plastic wrappers. In this authentic STEM activity, the students were (and continue to be) involved in environmentally friendly upcycling through the production and sale of recycled plastic pencil cases.

TEACHER NARRATIVE 6.2: Debbie Zaccias encouraged her students at Bumayong Lutheran Secondary School at Lae in Morobe Province to build a sawdust stove to benefit their school and community. Through this activity, they gained insights into how waste materials can be put to productive use. They also gained new skills and understandings in STEM.

TEACHER NARRATIVE 6.3: Gideon Aku used excursions and model building in teaching his Grade 12 Physics students at Utmei Secondary School in Rabaul, East New Britain Province. The students were taught the scientific principles of hydro-electricity, then visited a dam and power station before collaboratively building a model to demonstrate their understanding.

TEACHER NARRATIVE 6.4: Anastasia Paraka led her Grade 10 students at Mount Hagen Secondary School through the process of making herbal soap using medicinal plants. They gained hands-on knowledge of Indigenous plants from a number of sources including interviews with locals. She adopted a very specific focus on medicinal plants from the Western Highlands Province thus adding to the authenticity of the students' task.

TEACHER NARRATIVE 6.5: Koni Wamzii N'Drihin asked his Grade 11 students at Wawin National School of Excellence in Morobe Province to address the issue of waste management at their boarding school. They developed the "trashformer", an electrical transformer made from recycled metal and, in doing so, tackled a pressing local health and environmental issue.

Based on a comprehensive literature review, Herrington and Oliver (2000) described a pedagogical framework for authentic learning, identifying nine design elements applied in the Review section of this chapter. One offers that an authentic task or project can be integrated and applied across different subject areas and lead beyond domain-specific outcomes. This makes AL a good fit for an integrated STEM approach, as students engage in problem-solving involving multiple disciplines through the project (Sanders 2009). Students learn to investigate problems using skills and content knowledge from various disciplines requiring more than textbook formulas or rationales to solve (Tan et al. 2023).

The integrated STEM movement in schools has given increased attention to multidisciplinary and transdisciplinary approaches. By undertaking authentic real-world problems or projects, students apply knowledge and skills from two or more disciplines, helping them acquire knowledge, attitudes, and skills that develop their understanding of the characteristic features of STEM subjects (Bryan et al. 2015) and their STEM literacy (Williams 2017). The following five narratives provide insights into how teachers in PNG used a transdisciplinary approach to STEM integration and designed and implemented authentic activities in their local school and community contexts. Through their work, each teacher has brought disciplinary knowledge into play when addressing local, authentic issues and conditions.

6.2 Teacher Narrative 6.1

Paula Niandros Polume

6.2.1 Authentic Learning in STEM Through Recycling Waste Materials

I designed an authentic STEM project at ECOM (Evangelical Church of Manus) Secondary School in which my Grade 9 students made pencil cases, purses, and bags from recycled plastic. It ignited their interest in Mathematics, particularly in geometrical concepts relating to patterns and tessellation. It also showed them a way to earn a living while making a difference in their community by recycling materials that would otherwise be thrown away. My students enjoyed the creative activity of selecting colours and designs to make original products.

This project was authentic because it was sympathetic to the traditional weaving methods that our students know and love and provided a practical application of the geometry I was teaching in class. The resources used are familiar and readily available—almost every household in the township of Lorengau (Manus Island) buys and uses groceries wrapped in plastic and the disposal of these resources is an ongoing problem. The current practice of dumping plastics in the sea is unsustainable because of the environmental damage it causes. Finally, it allows students to become familiar with skills and processes which may help them to earn a living or supplement their income in the future by upcycling waste materials.

6.2.1.1 Rationale

Through integrating STEM, students learn to build models and apply a wide range of skills to solve everyday problems. They become creative and critical thinkers and authentic learners. They collaborate and communicate to find solutions or strategies for improvements. To help achieve this outcome, I try to engage my Grade 9 students in creating art with Mathematics and, along the way, integrating this with Design & Technology and Business Studies. Table 6.1 briefly introduces the problem statement, the connections to the curriculum and global goals, the pedagogy I chose, and my reflection on the outcomes of the project.

The project integrated different aspects of STEM (see Table 6.1). It also connected with the PNG curriculum in Grade 9 in four junior secondary subject areas:

- Business Studies: 9.1 Satisfying Needs and Wants (production and enterprise) (PNGDOE 2008a).
- Design Technology 9.1 Design and Technology Principles and Practice (Craft production) (PNGDOE 2006a).
- Mathematics: 9.1 Mathematics in our community Measurement, Geometry (Polygons and Tessellations) (PNGDOE 2006c).
- Science: 9.1 Working Scientifically—Measurement (PNGDOE 2006d).

6.2.1.2 Implementation

This project was implemented following the Teaching and Learning STEM Project-Based Learning (PBL) model (Forbes et al. 2021) through five phases:

Phase 1 [Ask]: In our Geometry class, we identified the polygons used in woven baskets, bilums, mats, walls of traditional houses, and head-dresses. We also identified repeated patterns and the process of tessellation and discussed the properties of materials used for weaving. Directing this lesson to alternate and possible materials to be used if we were to weave crafts, we discussed the use of plastics used as grocery wrappings and the daily rate of consumption.

Students were then asked to investigate what they could do about the management of waste plastic. They conducted online searches on waste management. I wanted them to understand how they could adapt and practise waste management in a positive way that could benefit them to create crafts to sell for home decoration or for personal use. Upon their discovery and observation of the use and care of used plastics, I suggested we make recycled plastic pencil cases, purses, or bags. The sample shown to the students motivated them to pursue this project.

Phase 2 [Imagine]: After seeing the sample, the students were organised into groups of six to plan the project. They were tasked with deciding on the size and design of their product, what plastic wrappers they needed to collect and to check on the prices in the local shops for thread, sewing needles and zippers. They made a budget plan, decided on the amount of money to be contributed and set deadlines for contributions, purchase of tools and collection of materials. They submitted their plans for checking.

Phase 3 [Plan]: Upon collection of the plastic wrappers and tools, the students decided which plastic wrappers they would use (sorted and selected by number, colour, size, the age of plastic, and durability). They then checked their designs to see if any adjustments were needed.

Phase 4 [Create]: I then invited my Aunty Kutan to the classroom to teach the students how to weave. The students observed how they could create their own pencil case, purse, or bag. They asked questions and made notes of solutions. After Aunty's class, I related her lesson back to Geometry, to explain how tessellation is part of weaving (Fig. 6.1). These are the mathematical processes that students would continue to use until their projects are completed.

Phase 5 [Improve]: Students gave feedback on their effort in completing this project. They also reflected on how they saw this project

Table 6.1 Narrative 6.1 project summary

Problem	Connections		Pedagogy	Reflection
Prevent environmental damage by upcycling plastic waste.	*Curriculum Connections*		STEM-PBL (Forbes et al. 2021)	Students recycled waste plastic to make pencil cases and other products using geometry and traditional weaving methods
	Grade 9 PNG Syllabuses: Maths (Geometry), Science, Design and Technology and Business studies.			
	SDGs (UN, n.d.)			
	1	No Poverty		
	4	Quality Education		
	8	Decent Work and Economic Growth		
	11	Sustainable Cities and Communities,		
	12	Responsible Consumption and Production		
	14	Life Below Water		

Fig. 6.1 Experiments with tessellation

as a change, or advocacy programme for waste management especially with managing plastic waste in our community. They have also contemplated how they disposed of their used plastics in the past. Having seen this as a way to earn a living, they discussed ways they could improve the current situation and were encouraged to think about how else they might better advertise and sell their products.

6.2.1.3 Outcomes

Herrington and Oliver (2000) believed that, for learning to be authentic, it has to have: an authentic context (a local issue), an authentic task (a product to be made and resources to be used), the presence of expert performances (from my Aunty), multiple perspectives (seeing the task as mathematics, craft, and environmental advocacy), collaboration (working together), reflection, articulation, metacognitive support, and authentic assessment. Using the rule, this simple task of using recycled materials to make a woven product was genuinely authentic (Fig. 6.2).

The five phases suggested by Forbes et al. (2021) were followed successfully to achieve outcomes that the students were happy with.

Fig. 6.2 Final woven products using recycled materials

Fig. 6.3 SDGs met through authentic learning at Lorengau (Manus Island) using waste materials

They appreciated using Mathematical concepts in creative arts and taking note of other subjects and goals integrated into this project. Most importantly, they learned to be advocates of environmentally safe and friendly practices that are sustainable and support self-reliance for productive citizens.

6.2.1.4 Reflection

I have conducted this activity only as a practical hands-on lesson to make Mathematics to be a fun subject to learn and not to be seen as a boring or content-driven and difficult subject.

Despite the activity not being part of the students' assessment, they were learning the curriculum content through their engagement with an authentic activity using recycling materials to make traditional woven products (bilums). This project aligns with the SDGs (Fig. 6.3). This includes SDG 1 [No poverty] and 8 [Decent work and economic growth] by finding ways to earn a living by selling bilums. It also aligns with SDG 11 [Sustainable Cities and Communities] and 12 [Responsible Consumption and Production] because it looked for ways to manage common plastic waste through purposeful recycling. The project, by potentially reducing waste being dumped in the ocean, also has a role to play in achieving the SDG 14 [Life below water].

This project shows how a simple classroom task can serve several learning and community goals. Students learnt about much more than weaving and created more than simple products. The sum was greater than its parts.

6.2.1.5 Postscript

I have continued and extended this task. The recycled pencil cases have become an Art and MAL (Making a Living) activity for Grades 7 and 8 students in ECOM Secondary schools. It is popular with the mothers who weave and sell at the markets. In February 2024, the Australian High Commission PNG media team featured this project on the International Day of Women and Girls in STEM (Australia Awards PNG 2024).

6.3 Teacher Narrative 6.2

Debbie Zaccias

6.3.1 Authentic Learning in STEM Through Construction of a Sawdust Stove

My school is Bumayong Lutheran Secondary School in Lae (Morobe Province). The school and the surrounding community usually cook outdoors because of ongoing electricity supply issues and the excessive cost of cooking fuels such as gas and kerosene (Fig. 6.4).

I wanted my students to build a more efficient means of cooking for the community. The most obvious choice was a sawdust stove. Using sawdust as fuel makes use of a waste by-product from the local timber industry. In this STEM project, Grade 12 students designed and collaboratively constructed durable sawdust stoves from sustainable materials. They made use of integrated STEM concepts learnt in Science and Mathematics and used tools and techniques from Design and Technology. This authentic hands-on project benefited student learning as well as the community. Table 6.2 introduces the problem statement, the connections to the curriculum and global goals, the pedagogy I chose and a brief reflection on the project.

6.3.1.1 Rationale

This STEM project began with the question, 'Why build a sawdust stove?' A simple sawdust stove can contribute to meeting five SDGs (Fig. 6.5). First, it provides affordable, reliable and sustainable energy [SDG 7]. Second, having properly cooked food is essential to good health and well-being [SDG 3] and reduces inequalities [SDG 10]. Third, using sawdust (an unwanted by-product of the timber industry) as fuel for stoves shows responsible consumption of dwindling resources [SDG 12]. Finally, it contributes to Quality Education [SDG 4] by engaging students in an authentic hands-on activity which meets both curricular and social needs.

There are several timber company outlets near the school that sell surplus timber waste (sawdust) at a reasonable price. The choice of sawdust as the fuel for the stove is based, however, on more than its availability as a waste product. The alternative of firewood is expensive—a bundle costs K5–K10 (approx. 1.39–2.78 USD$) (depending on the size of the bundle). With open-fire cooking, heat is easily lost to the surrounding environment which means that more firewood needs to be used. In addition, the smoke that the wood releases as it burns is a mixture of many different types of harmful gases including carbon dioxide (CO_2), carbon monoxide (CO), oxides of nitrogen (NOx) and volatile organic compounds (VOCs) which can be harmful to humans as well as contributing to the greenhouse effect. Therefore, constructing a sawdust stove which reduces the need for timber,

Fig. 6.4 Typical outdoor cooking methods in PNG

Table 6.2 Narrative 6.2 project summary

Problem	Connections		Pedagogy	Reflection
Electricity supply issues compounded with rising costs for other cooking fuels encourage inefficient and environmentally costly wood-burning outdoor cooking.	*Curriculum Connections*		STEM-PBL (Forbes et al. 2021)	While still underway, it is anticipated that the sawdust stoves will develop students 21st-Century skills while simultaneously teaching the necessary curriculum. It will also promote a clean, efficient and affordable cooking method for the local community
	Grade 12 Science (Physics and Chemistry), Maths, Design & Technology (Computer Studies, Metal Technology, Engineering)			
	SDGs (UN, n.d.)			
	3	Good Health and Well-Being		
	4	Quality Education		
	7	Affordable and Clean Energy		
	10	Reduced Inequalities		
	12	Responsible Consumption and Production		

Fig. 6.5 SDGs met through authentic learning in constructing a sawdust stove

and produces less smoke would provide a better solution.

The sawdust stove is durable as the metal plate used is an exceptionally good thermal conductor. Constructing sawdust stoves provides a hands-on experience for students to utilise and integrate STEM concepts as well as providing an efficient, sustainable, and affordable cooking solution for the community. Engaging students in STEM means discussing real-world problems that they want to solve rather than focusing on careers in STEM (Education Council, 2018).

According to Vasquez (2014), all STEM learning has one thing in common—it gives students opportunities to apply the skills and knowledge they have learned. Putting students into groups to work on the project reduces the time commitment and allows for a more even distribution of materials. Working in groups also allows for collaboration of ideas, innovation, and creativity by the students. By undertaking real-world problems or projects, students apply knowledge and skills from two or more disciplines and help to shape their learning experience (Vasquez et al. 2013).

This sawdust stove project integrates STEM concepts and makes meaningful connections with the PNG curriculum in Grades 11 and 12 as follows:

- Design & Technology (Computer Studies)—11.5 Desktop Publishing: Drawing and working with graphic objects (PNGDOE 2008h).
- Design & Technology and Engineering Design (PNGDOE 2008e).
- Mathematics (Advanced Mathematics) 11.1 Number and Application: Ratio & Proportion, Units of Measurement, (Geometry) Congruency, similarity, and construction (PNGDOE 2008b).
- Practical Skills (Practical Skills Technology, Metal Technology (PNGDOE 2008g).
- Upper Secondary Applied Science 11.3 Alternative Energy Sources (PNGDOE 2008c).
- Upper Secondary Chemistry 11.3 Types of Chemical Reaction Combustion Reaction (PNGDOE 2008d).
- Upper Secondary Physics: 12.2 Temperature & Heat. Temperature, Thermal Expansion, Heat & Heat Transfer (PNGDOE 2008i).

This integrated STEM project relies on all STEM disciplines. This should lead students to recognise that none of the disciplines can be omitted without leaving a conceptual gap (Lyons 2018).

6.3.1.2 Implementation

This authentic sawdust stove project was completed by senior Science students over a period of three weeks. I set four milestones for students to reach:

1. Planning and collection of materials, tools, and equipment to be used. Liaising with industry partner (welder) and school administration for funding (~3 days).
2. Planning and designing the stove (~3 days). Collaborating in teams.
3. Constructing, finishing and testing the stove (~6 days).
4. Reflecting on the project and creating a photo journal of the project (~3 days).

While the sawdust stove task represents an authentic activity, the pedagogy adopted was influenced by the five phases of the STEM-PBL approach (Forbes et al. 2021).

Phase 1 [Ask]. Students will decide the dimensions of the stove they will build and determine the volume of the sawdust required to fill up the hollow stove. They will list and cost the materials needed. This phase may also include liaising with the school administration for assistance in purchasing materials and requesting assistance from the welder employed by the school. Capraro et al. (2013) state that many schools that adopt STEM activities also create a broader self-sustaining learning community (PLC).

Phase 2 [Imagine]. This project will be carried out using the following materials:

1. Metal plate off-cuts—to be used as the base for the sawdust stove which will be welded together.
2. Tape measures—to measure the frame of the sawdust stove.
3. Welding machines and welding rods—to seal the cut-out metal plates together to make the frame of the stove.
4. Grinding machines and cutting wheels—to be used together to cut the metal plates into desired shapes for framing.
5. Chisels—to clear off rough edges from the metal.
6. Hollow PVC pipes—to be placed in the centre of the stove to allow ventilation and the flow of oxygen.
7. Welding helmets and safety boots—to be worn during the construction of the stove.

Metal sheets are used for the frame of the sawdust stove. Metals have very high thermal conductivity; thus, more heat is stored and less is lost. Fine sawdust is used in place of chunks of firewood because it yields a very clean heat and a slight flame. This is because the operation is by pyrolysis in excess air. As a result, the gases produce very little smoke.

Phase 3 [Plan]. Students will design the sawdust stove using their skills in Computer Studies (PGDOE, 2008h). They drew and worked with graphic objects using computer software such as Tinkercad and Adobe Illustrator. This activity focuses on creating a product or an artefact by using problem-based and inquiry-based learning depending on the depth of the driving question (Albion 2015).

Phase 4 [Create]. Metal sheets are cut by the cutting wheels using the grinding machine into eight identical slabs of 120mm by 300mm. The base has to be an octagonal metal cut shape with a sawdust feed inlet of 40mm by 300mm. All the metal slabs are then welded together by the welding machine using welding rods. This process will be carried out by the students with the assistance of an industrial partner (welder).

A hollow PVC pipe is then placed at the centre of the constructed stove and a certain volume of sawdust fills the metal casing. It is compacted enough so that there are no gaps in between. The metal sheets are excellent thermal conductors which store heat rather than giving off heat. The hollow gap is connected to the sawdust feed inlet to allow circulation of oxygen. This is a combustion reaction where the sawdust burns in the presence of oxygen, producing heat for cooking as well as light. The stove is ignited by using a gas lighter or match at the feed inlet causing heat to flow through the gap. The heat can be put out by placing a wet piece of cloth at the feed inlet, stopping oxygen from circulating.

The stove will be very efficient, sustainable, and reliable to the community (Fig. 6.6). It can also be economically beneficial to the school if more stoves can be constructed by the students and put out for sale to the surrounding community and the money goes back to the school.

Phase 5 [Improve]. After the completion of the project, students were asked to evaluate their work by identifying constraints, improving, and modifying their designs.

6.3.1.3 Outcomes

There are perceived and tangible outcomes of this project. It helped students to make use of 21st Century skills, particularly the ability to think critically and analytically and to develop higher-order thinking skills. It allowed collaboration, peer communication, problem-solving, and self-directed learning while expecting rigour (Capraro et al. 2013). Further, the production of the stoves brought a practical benefit to the community. With the ongoing electricity supply issues and the rising costs and scarcity of fuel, this project will provide the community with clean, affordable cooking.

The activity was both challenging and motivating. Future improvements and modifications could be made to the sawdust stove by connecting a thermoelectric generator to the feed inlet. This way, some of the heat produced could be used for cooking while some will be converted into electricity and used for light and charging mobile devices. The sawdust stove demonstrates how heat

Fig. 6.6 A partially completed sawdust stove

energy can be simultaneously used for multiple purposes.

6.3.1.4 Reflection

My school, along with others in Lae, has been experiencing low academic performances as well as some outbreaks of violence in the school. Creating such authentic activities may help students change their attitude to schooling, improve their thinking skills, and keep them busy with something educational.

A challenge to be faced in introducing projects such as this is working against competing deadlines. The school has schedules for both academic and extra-curricular activities. Projects such as this consume considerable time with many students involved but the outcomes serve multiple purposes.

This project fulfils some SDGs as well as connecting and integrating STEM concepts. Many ongoing issues in PNG can be solved if we start allowing students to think more critically, creatively, and collaboratively and to be more innovative in whatever they do. This can only be achieved if we integrate STEM concepts into our classroom teaching and learning. Authentic learning experiences consolidate students' learning in the classroom. It enhances real-world applicability and helps prepare students for post-secondary education, with an emphasis on making connections to what STEM professionals do in their job (Capraro et al. 2013).

6.4 Teacher Narrative 6.3

Gideon Aku

6.4.1 Authentic Learning in STEM from Excursions and Model Building

This narrative focuses on the connection between learning and the real world via excursion and site visits to the industrial areas and research institutions in the East New Britain Province (ENBP) where I am teaching. Exposing students to the real world gives them an opportunity to learn from experts working in the field as well as confirming the application of what they are learning in the classroom. The learning from excursions comes back to the classroom. My Grade 12 Physics students translated an excursion into a model-building exercise to demonstrate their understanding of the processes they had witnessed first-hand.

6.4.1.1 Rationale

A positive aspect about the excursion or real-world experience is that the students were learning from the experts in the field. The students see what the people are doing, and they can make good comparisons with what they learn in the classroom. They are making real-world connections from the classroom, broadening their horizons and experience of life. I drew on the principles of constructivism, social constructivism, and

Table 6.3 Narrative 6.3 project summary

Problem	Connections		Pedagogy	Reflection
Students needed opportunities to see connections between the curriculum being taught and real-world applications	*Curriculum Connections*		Constructivism (building knowledge from experience). Social constructivism (working in groups and learning with others). Constructionism (learning through model building).	By exposing the student to authentic environments based on STEM, they were able to collaboratively create a working mini-hydro turbine model.
	Grade 12 Chemistry, Biology and Physics.			
	SDGs (UN, n.d.)			
	4	Quality Education		
	9	Industry, Innovation and Infrastructure		

constructionism to design the different elements of their learning experience. Table 6.3 summarises the problem statement, the connections to the curriculum and global goals, the pedagogy I chose and a brief reflection on the project.

6.4.1.2 Implementation—The Excursion

This year, the Science department at Utmei Secondary School in ENBP organised three excursions for the Grade 12 Science students to visit industries and institutions within the province as part of their learning outside the classroom. The first two excursions, for Chemistry and Biology students respectively, are described briefly. The third, for Physics students, is described in greater detail and includes an extension tasks completed at school.*Grade 12 Chemistry Students' Excursion*: The Grade 12 Chemistry students visited one of the local soap production companies. They also visited Oil Palm Company to learn how the oil is extracted. The students made soap in our science lab with the help of their Chemistry teacher (see Fig. 6.7). Students had to work in groups to collect necessary equipment, chemicals, and ingredients for saponification. They learned something (such as making their own soap) that can be useful, and that knowledge can be sustained after leaving school. This has a direct connection to Unit 12.5 (Natural Resources and Chemical Industries in Papua New Guinea) in the PNG Chemistry Curriculum (PNGDOE 2008d). (See also Narrative 6.4 for a student activity in soap-making.)

Grade 12 Chemistry Students' Excursion: The Grade 12 Chemistry students visited one of the local soap production companies. They also visited Oil Palm Company to learn how the oil is extracted. The students made soap in our science lab with the help of their Chemistry teacher (see Fig. 6.7). Students had to work in groups to collect necessary equipment, chemicals, and ingredients for saponification. They learned something (such as making their own soap) that can be useful, and that knowledge can be sustained after leaving school. This has a direct connection to Unit 12.5 (Natural Resources and Chemical Industries in Papua New Guinea) in the PNG Chemistry Curriculum (PNGDOE 2008d). (See also Narrative 6.4 for a student activity in soap-making.)

Grade 12 Biology Students' Excursion: For the Grade 12 Biology students' excursion, I worked closely with the Grade 12 Biology teacher to organise the excursion. Before the excursion, the students were given a project on cloning, budding, and crafting. Their project was based on cocoa seedlings because our province is known for producing the most cocoa in the country. We decided to expose the students to the real world, so we wrote to the National Agriculture Research Institute (NARI) in East New Britain Province. We also wrote to the Cocoa Research Institute (CRI) in ENBP. Our requests were accepted, so we took the students down and they learned a great deal from scientists and local experts in the field (see Fig. 6.8).

Grade 12 Physics students' Excursion: The Grade 12 Physics students visited the Warangoi Hydro dam and powerhouse (PNG Power Property), the only Power Hydro dam in the East New Britain Province. The dam is seven kilometres

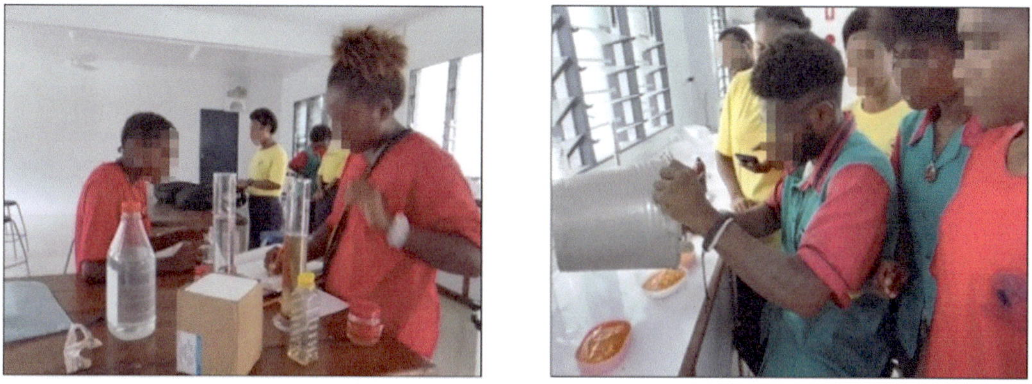

Fig. 6.7 Grade 12 Chemistry students making soap

Fig. 6.8 Grade 12 Biology students learning from the NARI and CRI experts

away from the powerhouse, so the students had to visit both to fully understand the concept of how electricity is produced. Understanding the function of each discrete facility was important, and provided the impetus for their later model building.

Inside the powerhouse, the students learned from what they saw and had the opportunity to learn from the experts (electrical engineers). Students toured the whole building with assistance from the workers in the powerhouse. As they walked around, the workers explained basic concepts of power generation as well as other operations in the powerhouse. The students had the opportunity to ask questions about power generation. Students learned a great deal from the real world apart from what they learn in the classroom.

6.4.1.3 Implementation—Model Building

To assess my students' understanding, they were asked to build a workable mini-hydro model based on their classroom learning and their excursion. They were also asked to present and demonstrate their working models at a Science Fair.

Building the model collaboratively was not only part of the students' assessment, but also part of their learning. Hallström and Schönborn (2019) argued that "the ability to create, use/apply, evaluate and revise models are necessary skills for acquiring an in-depth understanding of both technological development processes and scientific practice and are a core component for pursuing authentic learning in technology, mathematics and science classrooms" (p. 3).

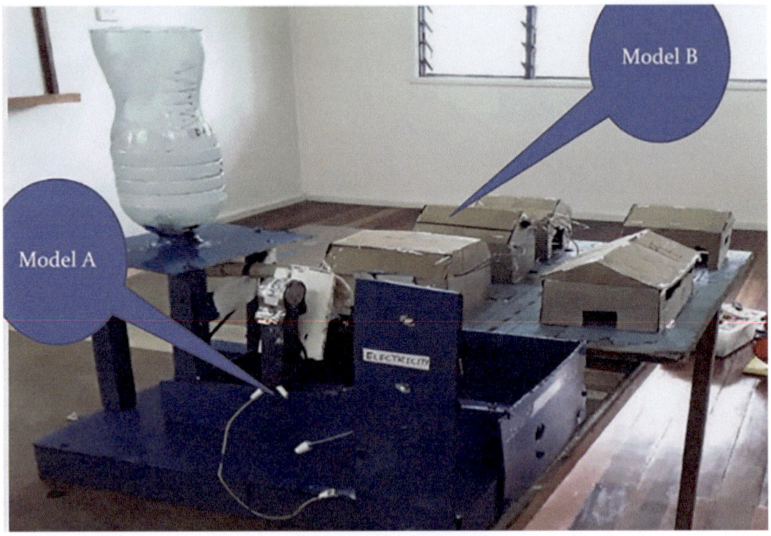

Fig. 6.9 Final model combining Model A and B

Building the model integrated different disciplines of STEM (see Fig. 6.9) and connected with the PNG curriculum in Grade 12 in three subject areas:

- Design and Technology: 12.1 Practical Skills Technologies unit (PNGDOE, 2008e).
- Mathematics: 12.1. Measurement (PNGDOE, 2006a).
- Physics: 11.5 Principles of Physics & 12.4 Electromagnetism (PNGDOE, 2008h).

I divided my students into two groups (five students per group). Group A was tasked to build the mini-hydro representing the section from the water inlet (dam) to the powerhouse and spinning turbine (Model A). Group B was tasked to build a model showing the transmission and distribution of the electric power produced from hydro (Model B). The duration of the project was two weeks with the students instructed to also use their free time to work on their models. They were encouraged to be creative and use whatever materials were available to them.

6.4.1.4 Outcomes

The students finalised their mini-hydro models in the allotted time. The two groups combined their projects to create a complete working mini-hydro

Fig. 6.10 SDGs met through excursions and model building

model. For Model A, the students used a motor and fan as the turbine to produce the current.

As planned, the students presented and demonstrated their models at the school's Science Fair. This event also included science games like building a tower and building bridge challenges were included in the programme as fun. The presentations were done in groups and were assessed using set rubrics. The presentations were impressive; students shared their knowledge with the audience through a PowerPoint presentation on the hydro dam and powerhouse followed by a demonstration of the combined mini-hydro model.

6.4.1.5 Reflection

It is useful to reflect on this activity in terms of its contribution to the SDGs, particularly SDG 9 [Industry, innovation, and infrastructure]

(Fig. 6.10). Following this activity, I am now trying to look at teaching and learning from another angle, that is, where students are mostly engaged in hands-on activities and do things by themselves. When the students built their models, they shared their knowledge with their peers and consolidated their learning from the field. It has been suggested that models and modelling can "be construed as one of the core tools or 'languages' of an integrative STEM literacy" and "meaningfully link the STEM disciplines" together (Hallström and Schönborn 2019, p. 8). I believe that this shows the project as working towards SDG 4 [Quality Education].

Another reflection relates to how my school's Science department is developing stronger relationships with PNG Power, NARI, CRI, and local soap production and palm oil companies. This is so our science students can be given further opportunities to come out of the classroom and learn from the field and to connect the learning in the classroom to learning in the real world. Field trips and excursions give great opportunities to the students to expand their knowledge about the things they learn in the classroom.

The excursions and model building provided an engaging and mutually supportive way to teach the syllabus content in the sciences. I am convinced of the effectiveness of authentic learning in teaching STEM in PNG.

6.4.1.6 Postscript

My students have continued to take excursions to the dam and powerhouse. The knowledge gained about electromagnetism is consolidated with model building, for example, the Grade 12 Physics students are planning to build a mini-hydro in front of the Science Lab and to build a mini-hydro for themselves when they will return home. I am still trying my best to help both teachers and students to really understand the concept of STEM.

6.5 Teacher Narrative 6.4

Anastasia Paraka

6.5.1 Authentic Learning in STEM Through Making Soap from Traditional Medicinal Plants

My Grade 10 students at Hagen Park Secondary made herbal soap using traditional medicinal plants (WHO, 2009). Soap-making is an authentic learning experience because soap is essential for our health and well-being. The activity integrates all STEM disciplines (see also Narrative 6.3 which describes soap-making as a study in Grade 12 Chemistry). I guided my students through the STEM 5E model (an inquiry-based learning approach), allowing them to be creative through problem analysis and exploration. The underlying aim of this project was to empower young Papua New Guineans to embrace their local traditional knowledge of medicinal plants and use it to improve their lives and sustain their livelihood. To this end, the students were told to only research traditional medicinal plants from the Western Highlands Province. My students researched and conducted interviews to collect traditional local knowledge on the uses of the medicinal plants.

6.5.1.1 Rationale

Making soap using traditional medicinal plants can arouse students' curiosity and creativity and allow them to consolidate their traditional knowledge about medicinal plants. My students enjoyed this activity and gained valuable local knowledge.

Students were able to focus on the preservation of Indigenous knowledge and skills to create innovative designs and use materials from their local environment. As the world is changing, there is a great need for the younger generation in PNG to embrace available local knowledge and skills to solve real-world problems.

Students were also able to engage in entrepreneurship where soaps can be produced for sale. When using STEM-PBL, students will be empowered with lifelong skills such as asking questions, researching, investigating, solving problems, collaborating, and communicating. They will build confidence instead of focusing on the curriculum alone. They become critical thinkers, authentic learners, and innovators full of ideas to solve real-world problems. According to the Organisation for Economic Co-operation and Development (OECD) report, productive individuals should have a high level of technical skills and be independent, cooperative, and flexible (Hakkarainen et al. 2004).

Table 6.4 introduces the problem statement, the connections to the curriculum and global goals, the pedagogy I chose and a brief reflection on the project.

This Grade 10 soap-making project integrates three subject areas within the PNG SBC curriculum (PNG Education News 2021). This allows students to develop the following knowledge and skills.

- Design & Technology (Computing skills)—Unit 10.4: Data communication and hardware. Use software and multimedia tools to automate data to help with the calculation, analysis, and processing of data for efficiency (PNGDOE 2006a).
- Design & Technology (Construction Technology)—Unit 10.1: Construct projects using correct handling techniques of tools and machines in building operations. (PNGDOE 2006a).
- Mathematics—Unit 10.3: Geometry, measurement, and transformation (PNGDOE 2006b). Demonstrate geometric thinking and mathematical skills, use previous mathematics knowledge to investigate geometric relationships. Connect three-dimensional shapes, volumes, and weights. Compare and contrast various three-dimensional models' volumes by considering their shapes.
- Science—Unit 10.1: Scientific Tool and Technology & Unit 10.6: Matter and Energy (PNGDOE 2006d). Use scientific process during laboratory investigation. Connect various chemistry topics in one experiment including: acids-bases, chemical reactions, chemical change, physical change, and chemical equations.

Making herbal soap can be connected to four Sustainable Development Goals (SDGs) (see Fig. 6.11). The first is SDG 3 [Good Health and Well-Being] which aims to e*nsure healthy lives and promote well-being for all ages.* Soap will be used for personal hygiene and cleaning. The second goal is SDG 12 which aims to *ensure sustainable consumption and production* because goods can be produced to meet human needs and make use of readily available ingredients. The third is SDG 8 which aims to create *decent work and economic growth.* When people make their own soap, they can sustain themselves financially by selling the soap to make money encouraging fair trade and entrepreneurship. The fourth relates to SDG 4 [Quality Education] in that soap-making offers hands-on learning which supports both curricular and traditional knowledge.

6.5.1.2 Implementation

Students followed the 5E Inquiry-Based Model (Bybee 2010) to make herbal soap. The model encourages scientific inquiry.

Engagement: The first phase of the 5E framework is not where the teacher lectures, defines terms, or provides explanations. It is, rather, the place for students to brainstorm and research the topic (Duran and Duran 2004). I began by showing the students some online videos about making herbal soap to build interest and curiosity for the project. They saw how to use oils and how to make soaps in particular geometric shapes and sizes. After watching the videos and collecting information, my class discussed why various geometric shapes and sizes are available around the world. I asked them which geometric shapes they might use and how much soap they might fit into one mould. I invited a community elder to give a

Table 6.4 Narrative 6.4 project summary

Problem	Connections		Pedagogy	Reflection
Preservation of Indigenous knowledge to solve local real-world problems.	*Curriculum Connections*		5E Inquiry-Based Model (Bybee 2010).	Students engaged in soap-making. The incorporation of traditional medicinal plants and sustainable practices provides an entrepreneurial opportunity for students to sell unique healthy products to the community.
	Grade 10 Science, Maths, Design & Technology (Computer Studies. Construction Technology).			
	SDGs (UN, n.d.)			
	3	Good Health and Well-Being		
	4	Quality Education		
	8	Decent Work and Economic Growth		
	12	Responsible Consumption and Production		

Fig. 6.11 SDGs met through authentic learning project making herbal soap

talk on the importance of traditional medicinal plants. I also had my students conduct interviews with locals who were involved in soap-making. Students then researched the properties of ingredients used in soap-making including traditional medicinal plants.

The students had to work in groups of eight because of the large number of students in my class. Each student was asked to keep a journal to show their understanding of soap production. They began by recording their preferred shape for a piece of soap, its dimension (height, width, circumference, radius, volume, weight, surface area, diameter, and total surface) and total weight. They would eventually create a report in a booklet form and contribute to a display for the whole school.

Exploration: The students were asked to explore the chemical processes involved in making soap. The steps must be followed meticulously using careful calculations and adhering to safety rules for handling the science equipment and materials. The ingredients needed were:

- Oil/fat: cooking oil either used or new, coconut oil or olive oil. In this project, students were taught to make pure coconut oil.
- Lye: Sodium hydroxide (NaOH), also known as caustic soda.
- Water.
- Essential oil or fragrance (optional).
- Dried powdered traditional medicinal herbs and plants.

Explanation: I explained lab safety procedures to the students before they started making their soaps. They began by combining the ingredients including the traditional medicinal plants they had chosen. Each group's progress was monitored and support given as needed. Because soap-making requires knowledge and skills from Chemistry, Mathematics, and Technology, I invited the teachers of these subjects to participate in this project.

Elaboration: The students could do additional research on additives such as colourants, essential oils, and fragrances to use in their soaps. After learning how to make soaps students can make similar products such as dishwashing detergent from fruit or vegetable peels. This diversification helps students to be creative in STEM.

Students can research the impact that soap-making has on the environment and other related issues such as sustainability and fair trade. Furthermore, they can also consider the artistic skills and creativity involved as well as the money-making potential of the production of soap. Finally, research on the endangered medicinal plant species in the students' own locality could be collated into a digital book to showcase these plants.

Evaluation: Students will present their herbal soaps to their peers who will give feedback based on four factors emerging from the method of production (curing) and types of materials used.

- *Appearance*: Herbal soaps may have a speckled or textured appearance but must be consistent with no discolouration or defects.
- *Texture*: Herbal soaps must be smooth and even rather than rough with air pockets. Soaps made with more lye will feel more cleansing and foamier while those with a high oil percentage will be creamy.
- *Performance*: Herbal soaps should have cleansing properties. The soap should feel soft and moisturising rather dry or strong. The medicinal power of the soap depends on the properties of the plants used.
- *Scent*: The intensity of the scent depends on the fragrance and number of essential oils used. The scent of the oil should be pleasant. It will fade as it cures over time.

Students will record what they have learnt, what they think is important and what prior information they used to solve problems in their personal journals. These entries will be checked to ensure that they have understood the processes. From these evaluations, I will share what students had omitted to try while making soap. I will reiterate aspects of the STEM disciplines involved in soap-making by using relevant illustrations or diagrams. Doing this will help me know how much my students have progressed from their original ideas and understanding and how their knowledge has changed over time.

6.5.1.3 Outcomes

Soap-making is a creative, fun, and fruitful project with opportunities for authentic STEM learning and the development of transferable skills (see Fig. 6.12).

The specific lessons learnt by students during this project include safety, accuracy and precision, control of temperature, timing and sequence, colour and fragrance and handling of medicinal plants.

- *Safety*: The handling of chemicals and equipment needs to be followed carefully to avoid harm and danger to students. Lye or caustic soda is dangerous and requires students to wear protective gloves and goggles to avoid burns and injury.
- *Accuracy and precision*: There must be coordination and collaboration between group members, checking and rechecking to make sure the correct measurement for each ingredient has been made. If there is too little or too much of any ingredient, it will affect the soap's texture, performance, and scent.
- *Control of temperature:* Controlling the temperature is important because if the mixture is too hot or too cool, it will affect the texture and quality of the soap. Everyone in a team needs to concentrate fully when monitoring the temperature.
- *Timing and sequence*: The timing of each process must be in correct sequence so that temperature is maintained. Excessive speed or slowness will have a negative effect on the appearance of the soap.
- *Colour and fragrance*: Adding colour and fragrance will have a delicate effect on the soap, but too much will be overpowering. This can reduce the value and give it a dull appearance. It is better to calculate how much is required before adding it.
- *Medicinal plants*: There must be deep knowledge and understanding of the uses and benefits of the plants selected for use. This will come from local knowledge and online research.

Fig. 6.12 Coconut oil soap (variety of shapes and colours)

The students would benefit from interviewing knowledgeable members of the local community.

6.5.1.4 Reflection

The most significant benefits of making soap using traditional medicinal plants are:

- *Cultural heritage*: Indigenous knowledge of traditional medicinal plants will be preserved for future generations.
- *Natural ingredients*: Traditional medicinal plants are organic and rich in natural substances. The soap will be moist, soft, and gentle unlike chemically originated soap which may cause irritation or allergy.
- *Healing properties:* Traditional medicinal plants added to the soap will assist in healing some skin diseases such as psoriasis, eczema, and acne and relieve itchiness.
- *Aromatherapy:* The aroma of traditional medicinal plants has a strong effect on the way a person behaves so, when added to soap, it will create a stress-free relaxing feeling.
- *Eco-friendly:* As with the use of any renewable resource, using traditional medicinal plants will reduce the impact of environmental damage because fewer chemicals will be used to make the soap thus reducing the emission of CO_2 in the atmosphere.
- *Integrated STEM:* In Science, students learn about chemical properties of the ingredients such as lye, oils, and water and how they react together. In Mathematics, they learn the importance of measurement and accuracy. They also learn about ratio and percentage.
- *Creativity:* The makers have control over the scent, texture, and colour of the soap. They can create soaps of different shapes and sizes.
- *Entrepreneurship*: Students can start a business in making soap and selling it in the community.

These eight benefits provide evidence of the value of soap-making as an authentic STEM activity. Connecting the 5E model to the STEM disciplines allows students to engage in authentic hands-on activities. Working in groups requires collaboration and communication with peers. Students working in groups to make herbal soap using traditional medicinal plants worked together to ensure correct processes were followed.

6.5.1.5 Postscript

This enjoyable and creative project seems to be very simple but it has genuine depth in its connection to STEM disciplines, Indigenous knowledge (particular to the Western Highlands Province) and entrepreneurship. Individual soaps were created based on research conducted by students, making the product an authentic result of their investigations.

6.6 Teacher Narrative 6.5

Koni Wamzil N'Drihin

6.6.1 Authentic Learning in STEM Processing Waste Materials to Improve Power Supply

Trashformer is a new word, a combination of trash and transformer. It refers to the electrical transformers that my Grade 11 students made from trashed meat and fish cans from the school canteen, locally called "the mess". This STEM activity was aimed at educating students to value waste as a resource which can be converted into useful materials. In turn, it can foster deep learning and incubate ideas for economic growth. This activity provided an opportunity for authentic learning as it addressed a local waste management problem.

For many boarding schools like Wawin National School of Excellence in Papua New Guinea, canned fish and meat seem to be the main source of protein to feed the students at the school mess. What to do with the empty cans (which cannot be recycled) is a pressing problem at our school, so I designed an authentic STEM learning activity to address the problem.

6.6.1.1 Rationale

Many boarding schools in Papua New Guinea are yet to develop a policy to handle school waste from the classroom, the dormitories, and the school mess. Empty food cans have been a perennial problem for many schools. The preferred method is to dispose of them at the dump site (Fig. 6.13). The problem does not end there. Empty cans attract flies, create an eyesore around the school, and emit a bad odour affecting the nearby staff residences. They are often taken away and scattered by stray dogs.

As a school, it is only proper to add value to these cans so they can land elsewhere and be put to good use. One use of these cans is to study their properties and see if they are useful elsewhere and if they can be brought into the classroom as part of a STEM project. Fish and meat cans are made of 98.5 % steel or iron and 1.5 % tin. According to Al Ghoul et al. (2020), tinplating is coating steel or iron with tin to prevent it from corrosion. This is interesting as it already contains iron or steel, a ferromagnetic material. By putting these cans to a new purpose involved my students in learning about the properties of metals and electricity concepts by making functional transformers. In this, I adopted a constructionist pedagogy (Papert & Harel, 1991) as "the best way to ensure that knowledge is built in the learner is through the active construction of something shareable — a poem, program, model or idea" (Stager 2016, para. 5).

Table 6.5 briefly summarises the problem statement, the connections to the curriculum and global goals, the pedagogy I chose and a reflection of the project.

Building the Trashformer model integrated different disciplines of STEM (see Table 6.6) and is connected to the PNG curriculum in Grade 11 in three subject areas:

- Design and Technology (PNGDOE 2008e).
- Mathematics (PNGDOE 2006a).
- Physics (PNGDOE 2008i).

6.6.1.2 Implementation

Proper planning and gathering of required tools and materials must be done first before the Trashformers were constructed. My students followed six steps in the construction.

Step 1 *Collection*: Students make special requests to the mess cooks for storage of empty meat or fish cans after cooking and collect the empty cans at an arranged time (Fig. 6.14).

Step 2 *Washing and Drying*: Students request funds from the administration for detergents or chemicals to wash the empty cans collected. The students use the detergents/chemicals to wash the cans before allowing them to dry.

Step 3 *Cutting, flattening and shaping*: With a tin snip or electric drill, the empty cans are cut into flat thin sheets. Using a hammer or electric hammer, dented or grooved sheets are flattened smoothly. With a cutting head on an electric hammer, the sheets can be cut into I and E shapes.

Fig. 6.13 School dump at NSoE Wawin

Table 6.5 Narrative 6.5 project summary

Problem	Connections		Pedagogy	Reflection
Increasing waste from the school mess not disposed of in an environmentally friendly way.	*Curriculum Connections*		Constructionism (learning through model building).	Students created "trashformers", electrical transformers from waste cans of meat from the school mess. This increased students' sense of environmental responsibility at their school and provided an authentic learning experience.
	Grade 11 Maths, Physics and Design & Technology.			
	SDGs (UN, n.d.)			
	3	Good Health and Well-Being		
	4	Quality Education		
	7	Affordable and Clean Energy		
	11	Sustainable Cities and Communities,		
	12	Responsible Consumption and Production		

Fig. 6.14 The meat cans available for trashformer construction

Step 4 *Drill and construction:* The cut-out I and E-shaped metal sheets are gathered to be drilled with an electric drill (see Fig. 6.15). The drilled sheets are then assembled to construct the transformer frame. The constructed transformer frame is insulated using insulated tape or paper.

Fig. 6.15 Student constructing a trashformer

Step 5 *Calculation of turns and coiling:* Students choose the power rating for the transformer and what type of transformer (step-up or step-down) they wish to build. Students use the transformer equation to calculate the number of turns required based on the desired power rating. This desired transformer rating needs to be checked and tested by an electrical professional. Coiling can commence after this has been done.

Step 6 *Testing*: Under the supervision of a qualified electrician, the Trashformer can now be tested to see if it functions well (Fig. 6.16). Students use instruments such as a voltmeter and an ammeter to confirm their input and output values of current and voltage and adjust where necessary.

The assessment for the activity was both formative and summative. It was customised to suit each step of the process. It is summarised in Table 6.6.

6.6.1.3 Outcomes

The students completed the construction of the Trashformer despite challenges that could have affected their progress. Those challenges are as follows:

- *Power supply*: The growth of an industry is propelled by having an adequate supply of electricity without unnecessary blackouts. In PNG, power blackouts are very common. Wawin NSoE has a supply pattern of 3 days power on and 4 days power off. To overcome this problem the school has a standby generator, but its operation depends on fuel which is very costly. This means that the best time for doing practical activities in the afternoon is almost always missed.
- *Tools*: A common phrase is a *'carpenter without tools cannot build a house.'* Likewise, science students without the tools for projects or practical work cannot fully comprehend scientific concepts hence lack scientific knowledge and skills. Unlike our sister schools at secondary level, Wawin NSoE lacks specialised buildings which should be fully equipped with manual or electrically operated tools for students to use.
- *Administrative support:* The key to any development in the school is support from the administration. It is paramount for the school to transform learning and embrace authentic learning. This support can be financial or through the supply of materials. Wawin NSoE funds projects from its petty cash resources.

6.6.1.4 Reflection

There are three benefits of engaging the students in this project. The first is environmental cleanliness. The appearance of the school grounds, especially around the mess, will be improved. The bad odours and flies attracted by the dumping of the empty cans will be reduced. This authentic project will contribute to cleanliness by regularly removing the empty cans or prevent them being dumped in the first place.

The second benefit is in developing student responsibility both within and outside of school. The school must not be seen as only a place where students and teachers gather to learn and impart knowledge academically but also be seen as a place where students become good responsible citizens. This project allowed students to take responsibility for a problem in the school and to understand the value of recycling waste materials to improve our environment. In this, it aligns with several SDGs (Fig. 6.17). It satisfies

Fig. 6.16 A constructed "Trashformer"

Table 6.6 Assessment for the Trashformer project

Steps	Types of assessment (Performed by teacher/s)
1. Collection	Formative assessment: *Check progress and respond to student feedback*
2. Washing and drying	Formative assessment: *Observation and supervision*
3. Cutting, flattening, and shaping	
4. Drilling and construction	
5. Calculation of turns and coiling	Formative: *Observation and supervision* Summative:
6. Testing	• *Assignment (students show competency in calculating number of turns and confirm their result by the use of instruments.* • *Students make comments comparatively about the data collected and manipulated.*

SDG 3 [Good Health and Well-Being] by encouraging students to actively remedy a problem with negative impacts on their health and well-being. It also relates to SDG 11 [Sustainable cities and communities] and SDG 12 [Responsible Consumption and Production] because it looked for ways to manage community waste through purposeful recycling and maintain a healthy and habitable environment. The completed products, the transformers, can be used as a component in power distribution and therefore contributes to SDG 7 [Affordable and Clean Energy]. Above all, it related to SDG 4 [Quality Education] because it acted rather than just talked about an issue, teaching students how learning in STEM can be used to tackle problems in the real world.

The third and final benefit was in engaging the students in authentic activities to develop their talents and apply the knowledge they have gained in class. The vision of Wawin NSoE is to

Fig. 6.17 SDGs met through authentic learning through making a "trashformer"

produce students who are academically excellent and are good citizens. Engaging the students in a project such as Trashformer helps to achieve this vision, meeting both curriculum requirements and addressing an authentic local problem. The activity not only achieves the academic part of the vision but moulds students to be responsible citizens who can be productive when they return to their communities.

6.7 Review of Authentic Learning in STEM in PNG

Senka Henderson and Charisse Farr

The Authentic Learning narratives of the five teachers presented here (6.1–6.5) offer insights into how authentic learning is being implemented in PNG schools. While the pedagogical approaches and discipline integration varied among the teachers, each narrative showcased the teachers' ability to connect discipline knowledge to real-world issues and local contexts. That some of these projects are continuing indicates the esteem in which they are held within their school communities and how genuinely authentic they are in the context in which they were originally designed.

Herrington and Oliver's (2000) authentic learning framework can be used to analyse the teacher narratives presented in this chapter. The framework includes nine design elements that create a learning environment that fosters authentic learning. Those elements are:

1. An authentic context that reflects the way the knowledge will be used in real life
2. Authentic activities
3. Access to expert performances and modelling of processes
4. Multiple roles and perspectives
5. Collaboration
6. Reflection
7. Articulation
8. Coaching and scaffolding
9. Authentic assessment.

6.7.1 Herrington and Oliver's Framework

Below we discuss each of the elements of Herrington and Oliver's framework in the context of the authentic STEM activities presented in the teacher narratives in this chapter.

6.7.1.1 Authentic Context

An authentic context is the cornerstone of authentic learning providing the purpose and motivation for learning. It includes a physical environment (such as classroom, excursions, or out in the community) where the project is organised around a central context (Herrington & Herrington, 2006; Herrington & Oliver, 2000.)

In the teacher narratives, authentic context was achieved through dealing with real issues relating to the teachers' communities. Some addressed more than one issue revealing the complexity of the PNG environment and offering opportunities for differing aspects of STEM to be integrated in their possible solutions. Three narratives dealt with the context of waste materials. Paula Niandros Polume [6.1] found an educational use for plastic waste in her community, Debbie Zaccias [6.2] used sawdust, while Koni Wamzii N'Drihin [6.5] converted metal food cans to transformers. In each instance, a genuine community issue was addressed. Student awareness was raised about recycling and STEM knowledge was applied to reusing waste materials. Debbie and Koni also addressed issues of reliable sustainable energy. A number of other narratives in this book also make use of an authentic context to situate and give additional meaning to students' learning in STEM (see, e.g., Manasseh Kui Fena 7.3; Levesa Morumo 8.2; Abel P. Mambo 8.3).

6.7.1.2 Authentic Activities

The task that students perform is the most crucial part of the learning environment. Multiple research studies have focused on defining the characteristics of authentic activities and ten broad characteristics have been identified (e.g. Herrington and Herrington 2006; Herrington and Oliver 2000; Reeves et al. 2002). Thus, authentic activities should:

1. have real-world relevance,
2. be ill-defined, requiring students to define the tasks and sub-tasks needed to complete the activity,
3. comprise complex tasks to be investigated by students over a sustained period of time,
4. provide the opportunity for students to examine the task from different perspectives, using a variety of resources,
5. provide the opportunity to collaborate,
6. provide the opportunity to reflect,
7. be able to be integrated and applied across different subject areas and lead beyond domain-specific outcomes,
8. be seamlessly integrated with assessment,
9. create a product rather than an exercise or substeps preparing for something else, and
10. allow competing solutions and diversity of outcome.

All the teacher narratives of authentic learning in STEM feature students who took part in activities that had real-world relevance as evidenced in the conscious and deliberate references to the Sustainability Development Goals (see Appendix 1). For example, SDG 12 is concerned with responsible consumption and production. This, in different forms, was addressed in three further narratives [Paula 6.1; Debbie 6.2; and Anastasia 6.4].

While all five teacher narratives presented defined tasks for the students, the sub-tasks and resources were left undefined. This gave students the freedom to choose different materials, patterns, colours, and products (as it was for Anastasia's students in their creation of soaps in a variety of shapes, sizes, and textures). Students were also encouraged to be creative in repurposing and using whatever materials were available to them (as in the models made by Gideon Aku's students [6.3]). In most cases, the students had to work around sub-optimal classroom conditions and even endure not having electricity for most of the day. However, in these narratives, the students had to seize the opportunity that was born out of necessity, which is often the reality of situations in developing countries.

As mentioned earlier, authentic learning is the capacity to integrate tasks across different subject areas. The teacher narratives in this chapter provide several examples of subject integration, with most narratives showcasing STEM integration. However, some narratives go beyond STEM and incorporate other subjects. For example, Paula's students using recycled materials to make traditional woven products were learning STEM content as well as content from Business Studies to demonstrate how they can earn a living and supplement their income in the future.

6.7.1.3 Access to Expert Performances and the Modelling of Processes

An authentic learning environment needs to expose students to expert thinking. According to Herrington and Herrington (2006), for students to develop professional performance, they should be given "a model of how real practitioners behave in a real situation" (p. 5). In the teacher narratives, students had access to expert performance in two ways. First, students learned the craft of weaving from a community member (Paula's aunt) while Anastasia's students learnt about traditional medicinal plants from a local expert. Second, students learned from experts in the field during school excursions. Gideon's students learned STEM-related concepts when they worked with technicians, scientists, and electrical engineers during their visits to local companies (such as soap production and oil palm), research institutes (such as the National Agriculture Research Institute and Cocoa Research Institute), and the hydro dam and powerhouse. These students gained a new understanding of their own community along with the STEM knowledge they were acquiring.

6.7.1.4 Multiple Roles and Perspectives

An authentic learning environment should enable students to explore different perspectives. In the teacher narratives, multiple perspectives were provided within the learning environment in three distinct ways:

- students were required to work in small groups and each student brought their own perspectives;
- students adopted differing roles within a team and provided additional monitoring on critical processes;
- students presented their final products to class and community (during science school fairs), which enabled the expression of different points of views and peer evaluation.

6.7.1.5 Collaboration

The opportunity for students to collaborate and work in small groups is an important element of authentic problem-solving activities. An excellent example of collaboration can be seen in Gideon's teacher narrative. During the building of the mini-hydro model, students started their project by working in two smaller groups of five students. However, towards the end of the project, the groups were combined, and the final model was a product of the collaboration between all students. In support of the collaborative construction of knowledge, grades for a class presentation at the school's Science Fair were given for group work and not individually. Having students work in groups is sometimes a logistical necessity because of time and resource constraints.

6.7.1.6 Reflection

Reflection is an important element in finding the solution to an authentic task. Some researchers (Knights 1985) see reflection as a two-way process with the "aware attention" of another person (p. 85) or as both a process and a product (Collen 1996). In the teacher narratives in this chapter, there are restricted opportunities for students to reflect on their projects because of time pressures of the classroom and curriculum. However, Paula's students gave feedback to each other on their "products" (i.e., bilums, pencil cases and purses) and actively reflected on the "process" of their learning throughout the project.

6.7.1.7 Articulation

The process of articulation enables the formation, awareness, development, and refinement of students' thoughts (Herrington and Herrington 2006). An authentic learning environment should provide opportunities for students to work collaboratively, articulate their ideas and present and defend their arguments in front of an audience. In Gideon's narrative, students articulated their final products and understanding of STEM content knowledge in class in front of their peers and in front of wider audience at the school's science fair.

6.7.1.8 Coaching and Scaffolding

In authentic learning environments, coaching and scaffolding are mostly done by the teacher, although it can be provided by more able students or by other means. Gideon's students presenting their work to an audience of peers served the same purpose.

6.7.1.9 Authentic Assessment

The learning environment needs to provide opportunities for students to demonstrate their content knowledge, process, and product in collaboration with others. An authentic assessment also needs to be directly integrated with the activity and to provide appropriate marking criteria for scoring. In four narratives, teachers had designed summative assessment, and all initiated a regular schedule of formative assessment to encourage and guide students.

6.8 Final Thoughts

The teacher narratives in this chapter showcase the importance of integrated STEM in authentic learning in classroom teaching. While the approaches employed by teachers varied, all projects shared a common thread. That is, each addressed identified problems and challenges that students faced in their communities and provided practical and relevant learning opportunities for students.

The challenges, faced by students in developing countries such as PNG, are often vastly different from those in Western industrialised countries. For example, it may be difficult for students in PNG to access even basic classroom resources and materials, such as pencils and paper, let alone advanced digital technologies. In addition, students in PNG may have to navigate overcrowded classrooms and limited access to electricity, which can make it challenging to implement authentic learning experiences. Furthermore, many schools in PNG lack financial assistance for sourcing additional educational resources. Consequently, schools often rely on their own limited funds to finance school projects and secure necessary materials, using their own petty cash resources.

Despite these challenges, the teacher narratives demonstrate that teachers and students in PNG can engage in authentic learning experiences that address global issues, such as poverty, health, and waste management. Anastasia's students had to work in groups of eight because of limited resources to be shared in a large class. This shows that, with creativity and resourcefulness, it is possible to overcome some of the obstacles to authentic learning in developing countries. Through innovative solutions and creative use of materials, these students were able to learn and develop the skills they need to succeed in their future careers. It is important to recognise the unique perspectives and challenges of students in developing countries and to support their educational journeys through the implementation of authentic learning experiences.

References

Al Ghoul L, Abiad MG, Jammoul A, Matta J, El Darra N (2020) Zinc, aluminium, tin and Bis-phenol a in canned tuna fish commercialized in Lebanon and its human health risk assessment. Heliyon 6(9):e04995. https://doi.org/10.1016/j.heliyon.2020.e04995

Albion P (2015) Project-, problem-, and inquiry-based learning. In: Henderson M, Romeo G (eds) Teaching and digital technologies: Big issues and critical questions. Cambridge University Press, pp 240–252

Anker-Hansen J, Andreé M (2019) In pursuit of authenticity in science education. NorDiNa, Nordic Studies in Science Education 15(1):498–510. https://doi.org/10.5617/nordina.4723

Australia Awards PNG (2024, February 15) STEM teacher leads by example. https://www.australiaawardspng.org/2024/02/15/stem-teacher-leads-by-example/

Bryan L, Moore T, Johnson C, Roehrig G (2015) Integrated STEM education. In: Johnson C, Peters-Burton E, Moore T (eds) STEM roadmap: a framework for integration. Routledge, pp 23–37

Bybee RW (2010) Advancing STEM education: a 2020 vision. Technol Eng Teach 13(6):30–35
Capraro R, Capraro MM, Morgan J (2013) STEM project-based learning: an integrated Science, Technology, Engineering, and Mathematics (STEM) approach. Sense
Chopra A, Banerjee P (2022) Authentic learning for developing key competencies in sustainability: a review. J Posit Sch Psychol 6(3):9021–9034. https://journalppw.com/index.php/jpsp/article/view/5220
Collen A (1996) Reflection and metaphor in conversation. Educ Technol 36(1):54–55
Duran LB, Duran E (2004) The 5E instructional model: A learning cycle approach for inquiry-based science teaching. Sci Educ Rev 3(2):49–58
Forbes A, Chandra V, Pfeiffer L, Sheffield R (2021) STEM education in the primary school: a teacher's toolkit. Cambridge University Press
Hakkarainen K, Palonen T, Paavola S, Lehtinen E (2004) Communities of networked expertise: Professional and educational perspectives. Elsevier Scientific
Hallström J, Schönborn KJ (2019) Models and modelling for authentic STEM education: reinforcing the argument. Int J STEM Educ 6(22). https://doi.org/10.1186/s40594-019-0178-z
Herrington A, Herrington J (2006) What is an authentic learning environment? In: Herrington A, Herrington J (eds) Authentic learning environments in higher education. Information Science Publishing, pp 1–13
Herrington J, Oliver R (2000) An instructional design framework for authentic learning environments. Educ Technol Res Dev 48:23–48. https://doi.org/10.1007/BF02319856
Herrington J, Parker J (2013) Emerging technologies as cognitive tools for authentic learning. Br J Educ Technol 44(4):607–615. https://doi.org/10.1111/bjet.12048
Herrington J, Reeves TC, Oliver R (2014) Authentic learning environments. In: Spector J, Merrill M, Elen J, Bishop M (eds) Handbook of research on educational communications and technology. Springer. https://doi.org/10.1007/978-1-4614-3185-5_32
Knights S (1985) Reflection and learning: the importance of a listener. In: Boud D, Keogh R, Walker D (eds) Reflection: turning experience into learning. Kogan Page, pp 85–90
Lombard MM (2007). Authentic learning for the 21st Century: an overview. Educ Learn Initiat 1, 1–12. https://library.educause.edu/resources/2007/1/authentic-learning-for-the-21st-century-an-overview
Lyons, T. (2018). Helping students make sense of STEM. *Teaching Science: The Journal of the Australian Science Teachers Association*, *64*(3), 37–40. https://doi.org/10.3316/informit.841550904446404
Moeed A, Anderson D (2018) Learning through school science investigation: Teachers putting research into practice. Springer Singapore. https://doi.org/10.1007/978-981-13-1616-6
Nachtigall V, Shaffer D, Rummel N (2022) Stirring a secret sauce: A literature review on the conditions and effects of authentic learning. Educ Psychol Rev 34:1479–1516. https://doi.org/10.1007/s10648-022-09676-3
Palm T (2008) Performance assessment and authentic assessment: a conceptual analysis of the literature. Pract Assess, Res Eval 13(4):1–11. https://doi.org/10.7275/0qpc-ws45
Papert S, Harel I (1991) Constructionism. Ablex
PNGDOE (Papua New Guinea Department of Education) (2006a) Design and technology subject field lower secondary syllabuses. https://educationpng.gov.pg/Teachers/secondary/lower/syllabus-lower-secondary-design-technology.pdf
PNGDOE (Papua New Guinea Department of Education) (2006b) General mathematics upper secondary syllabus. https://www.education.gov.pg/syllabus/secondarysyllabus/subjectssyllabus/General%20Mathematics%20Grade%2011%20&%2012%20Syllabus.pdf
PNGDOE (Papua New Guinea Department of Education) (2006c) Mathematics lower secondary syllabus. https://educationpng.gov.pg/Teachers/secondary/lower/syllabus-lower-secondary-mathematics.pdf
PNGDOE (Papua New Guinea Department of Education) (2006d) Science lower secondary syllabus. https://educationpng.gov.pg/Teachers/secondary/lower/syllabus-lower-secondary-science.pdf
PNGDOE (Papua New Guinea Department of Education) (2008a) Business studies upper secondary syllabus. https://www.education.gov.pg/syllabus/secondarysyllabus/subjectssyllabus/Business%20Studies%20Grade%2011%20&12%20Syllabus.pdf
PNGDOE (Papua New Guinea Department of Education) (2008b) Advanced mathematics upper secondary teacher guide. https://www.education.gov.pg/syllabus/secondarysyllabus/subjectssyllabus/Advanced%20Mathematics%20Grade%2011%20&12%20Syllabus.pdf
PNGDOE (Papua New Guinea Department of Education) (2008c) Applied Science upper secondary syllabus. https://educationpng.gov.pg/Teachers/secondary/upper/syllabus-upper-secondary-applied-science.pdf
PNGDOE (Papua New Guinea Department of Education) (2008d) Chemistry upper secondary teacher guide. https://png.oer4pacific.org/id/eprint/324/1/syllabus-upper-secondary-chemistry.pdf
PNGDOE (Papua New Guinea Department of Education) (2008e) Design and technology upper secondary syllabus. https://www.educationpng.gov.pg/Teachers/secondary/upper/teachers-guide-upper-secondary-design-technology.pdf
PNGDOE (Papua New Guinea Department of Education) (2008f) Design and technology upper secondary teacher guide. https://www.educationpng.gov.pg/Teachers/secondary/upper/teachers-guide-upper-secondary-design-technology.pdf
PNGDOE (Papua New Guinea Department of Education) (2008g) Geography upper secondary syllabus. https://www.education.gov.pg/syllabus/secondarysyllabus/subjectssyllabus/Geography%20Grade%2011%20&%2012%20Syllabus.pdf

PNGDOE (Papua New Guinea Department of Education) (2008h) Information communication technologies and computer studies upper secondary teacher guide. https://educationpng.gov.pg/Teachers/secondary/upper/teachers-guide-upper-secondary-information-communication-technologies-computer-studies.pdf

PNGDOE (Papua New Guinea Department of Education) (2008i) Physics upper secondary syllabus. https://www.education.gov.pg/syllabus/secondarysyllabus/subjectssyllabus/Physics%20Grade%2011%20&%2012%20Syllabus.pdf

PNG Education News (Papua New Guinea Education News) (2021, May 22) When and why was standard based education or curriculum introduced in PNG? https://edu.pngfacts.com/standard-based-education/when-and-why-was-standard-based-education-or-curriculum-introduced-in-png

Reeves TC, Herrington J, Oliver R (2002) Authentic activities and online learning. In: Quality Conversations, Proceedings of the 25th HERDSA Annual Conference. Perth, Western Australia

Sanders M (2009) STEM, STEM education, STEMmania. Technol Teach 68(4):20–26

Stager G (2016) Seymour Papert (1928–2016). Nature 537:308. https://doi.org/10.1038/537308a

Tan AL, Ong YS, Ng YS, Tan JHJ (2023) STEM problem solving: Inquiry, concepts, and reasoning. Science & Education 32:381–397. https://doi.org/10.1007/s11191-021-00310-2

United Nations (n.d.) Sustainable development goals. https://www.un.org/sustainabledevelopment/sustainable-development-goals

Vasquez JA (2014) STEM: beyond the acronym (integrating science, technology, engineering and mathematics in teaching). Educ LeadShip 72(4), 10–16. https://qut.primo.exlibrisgroup.com/permalink/61QUT_INST/1fes5bt/cdi_gale_infotracacademiconefile_A394320322

Vasquez JA, Sneider C, Comer M (2013) STEM lesson essentials, Grades 3–8: integrating science, technology, engineering, and mathematics. Heinemann

WHO (World Health Organisation) (2009) Medicinal plants in Papua New Guinea: Information on 126 commonly used medicinal plants in Papua New Guinea. https://www.who.int/publications/i/item/9789290612490

Williams J (2017) The vocational goals of STEM education: Is that enough? Paper presented at PATT 2017. Philadelphia: Technology & Engineering Education – Fostering the Creativity of Youth Around the Globe July 10–14, 2017

Senka Henderson is an Associate Professor at QUT, a Senior Fellow of the Higher Education Academy (HEA) and an Associate Fellow (Indigenous) of the HEA. She is the Course Co-ordinator of the Master of Teaching (Secondary) programme. Senka teaches multiple curriculum, assessment, and pedagogy units in secondary studies across both undergraduate and postgraduate courses. She was involved in the case study research which underpins Part I. She contributed to Chap. 3 and led the writing of Chap. 4. Along with Charisse Farr, she wrote the Introduction to and Review of this chapter on *Authentic Learning*.

Charisse Farr is a Senior Lecturer of Secondary Mathematics Education at QUT. She teaches secondary Mathematics education, ways of thinking in STEM and STEM education, Calculus and pre-calculus and first year statistics across multiple disciplines. Charisse has over a decade of experience in integrating mathematical and statistical methods to understand real world scenarios and solving associated problems. Along with Senka Henderson, she wrote the Introduction to and Review of this chapter on *Authentic Learning*.

Paula Niandros Polume was, at time of writing, a Mathematics and ICT teacher at ECOM (Evangelical Church of Manus) Secondary School, Manus Island with almost 10 years of teaching experience. She has since been promoted to Subject Master of Mathematics, pending further advancement to Head of Mathematics. She advocates STEM Education by leading in-services, producing a resource book and supporting funding efforts. She has returned to Queensland University of Technology, through an AAPNG scholarship, to undertake a Master's degree in Education. Her contributions to this book are through this chapter *Authentic Learning* through recycling waste materials, and Chap. 8 *Project-based Learning* providing reliable access to water through the construction of Tippy Taps.

Debbie Zaccias is Mathematics and Physics teacher at Bumayong Lutheran Secondary School in Lae. She has eight years teaching experience. She contributed to this chapter *Authentic Learning* in STEM through the construction of sawdust stoves, a reliable and efficient means of re-using waste materials.

Gideon Aku at time of writing, was a Teacher of Mathematics and Physics at Utmei Secondary School, Rabul, in the East New Britain Province with 9 years' experience. He has since been promoted to Deputy Principal (Academic) and is now recognised as a leader in STEM Education in his province and trains teachers in very remote schools, frequently travelling long distances by boat. He contributed to this chapter, *Authentic Learning* with a project focussed on developing STEM knowledge through Excursions and Model Building.

Anastasia Paraka at time of writing, was the Head of Science at Mount Hagen Secondary School with 16 years teaching experience. She is now the Head of Agriculture at Kui Wamp Nga Secondary School, still within the Western Highlands Province). Anastasia has contributed to this chapter *Authentic Learning* teaching the process of making soap from Indigenous medicinal plants. Important elements of the project centre on empowering young Papua New Guineans to embrace local traditional knowledges and entrepreneurship.

Koni Wamzil N'Drihin is a Physics and Computer Studies Teacher at the Wawin National School of Excellence boarding school with over 15 years of teaching experience. He has contributed to this chapter *Authentic Learning* with a project focussing on processing waste materials to improve the transmission of power.

Design Thinking in STEM

7

Christopher N. Blundell, Andy Yeh, Douglas Hopkos, Lonah Tombele, Manasseh Kui Fena, and Raphael Ulka

Abstract

This chapter explores the use of design thinking in STEM Education. It provides insight into the application of design thinking in STEM teaching and learning by drawing on teachers' narratives about learning experiences that they designed. The chapter starts by providing an overview of design thinking and its usefulness to teachers and students. The four PNG teacher narratives provide rich examples of innovative, design-based learning experiences, aligned to the United Nations Sustainable Development Goals. The chapter concludes by reviewing how the teachers used design thinking to solve problems of practice, and highlights how students learned STEM through design.

Keywords

STEM · Design thinking · Teacher-led innovation · Design-based learning

7.1 What Is Design Thinking in STEM?

Christopher N. Blundell and Andy Yeh

Design thinking has its origins in how designers and engineers engage with and solve complex, human-centred problems. At its most fundamental level, design thinking involves understanding a problem by engaging with people's experiences of it (the problem definition phase) and then developing, testing and evaluating potential solutions in response (the solution development phase) (Brown and Katz 2009; Henriksen et al. 2020; Razzouk and Shute 2012).

Rationales for using design thinking approaches in education focus on its benefits to students and teachers. For students, engaging in design thinking creates opportunities to participate in problem-solving activities and develop a wide range of lower and higher order cognitions (Koh et al. 2015). This includes critical and creative thinking (Razzouk and Shute 2012), and

C. N. Blundell (✉)
School of Education, Queensland University of Technology, Kelvin Grove, QLD, Australia
e-mail: christopher.blundell@qut.edu.au

A. Yeh
School of Education, Queensland University of Technology, Kelvin Grove, QLD, Australia

D. Hopkos
Bishop Leo Arkfeld Secondary School, Wewak, Papua New Guinea

L. Tombele
Kilakila Secondary School, Port Moresby, Papua New Guinea

M. K. Fena
Wawin National School of Excellence, Lae, Papua New Guinea

R. Ulka
Gumine Secondary School, Kudiawa, Papua New Guinea

V. Chandra et al. (eds.), *Integrated STEM Education in Papua New Guinea*, Sustainable Development Goals Series, https://doi.org/10.1007/978-981-95-0828-0_7

Table 7.1 Common actions in design thinking models

Category	Problem definition phase		Solution development phase		
Design thinking models	Empathise, Discover	Define, Interpret	Ideate	Prototype, Experiment	Test and Evolve
Engineering design process models	Identify a need or problem	Research need or problem	Develop and select best possible solutions	Construct a prototype	Test and evaluate solution

Drawn from Blundell (2022)

generates opportunities to be empathetic (Retna 2016).

Observations of design practitioners' processes have resulted in researchers and professional associations developing multiple design thinking models (Razzouk and Shute 2012)—for an example of the range see Bobbe et al. (2016). The models typically fall into two categories: engineering design process models (based on the work of engineers) and design thinking models (based on the work of non-engineers). Table 7.1 offers a summary of existing models.

Although the language used by respective models differs, the essential actions are largely consistent. Design thinking involves a combination of critical thinking and creativity that draws upon and is confirmed by engaging with others (Henriksen et al. 2020). In this chapter, we will use the term "design thinking" to collectively describe both categories.

Design thinking has emerged in STEM Education as a human-centred approach to problem-solving (English et al. 2020) that is relevant to all fields of STEM (Li et al. 2019). For students, design thinking is positioned as a 21st-Century skill that is relevant to life and work in a complex world (Koh et al. 2015; Razzouk and Shute 2012). Engaging in design thinking in STEM Education is an authentic mode of learning because students are solving real-world problems (Bush et al. 2018).

For teachers, design thinking is positioned as an approach to addressing problems of practice (Henriksen et al. 2020; Retna 2016) and as a method for professional learning (Blundell 2022). It is for this reason that this project included opportunities for teachers to learn about and apply design thinking in their teaching practice.

English et al. (2020) used the term "design-based learning" to describe the teaching approach that engages students in learning about design as well as solving open-ended problems through designing innovative products or solutions. They further identified two forms of design-based learning as "learning about design" and "learning through design". Learning *about* design focuses on the theoretical concepts of design thinking and processes, while learning *through* design focuses on the development of disciplinary concepts (e.g., STEM) during the design of products or solutions. It is important to note that the two forms are not necessarily exclusive to each other in teaching practice as often teachers could employ both forms concurrently. In recognition of its use in industry and its value for students, design thinking is being incorporated in STEM curriculum documents. For example, design thinking is specifically incorporated as a core concept for creating solutions in the Australian Curriculum Technologies (ACARA 2023).

For teachers, design thinking proponents argue that, by its nature, teaching is a design process of understanding students' needs and developing responsive learning experiences (Billiar et al. 2014; Henriksen et al. 2020). Kim (2021) found that STEM teachers use design thinking to create new learning experiences and pedagogies, to repurpose or adapt technologies, and to redesign curriculum documents. Teachers can also use design thinking as a professional learning and development process to engage with and respond to stakeholders, and then to iteratively develop and test new approaches (Blundell 2022). Yurkofsky et al. (2020) argued that design thinking is a viable way to facilitate educational change at classroom, team, school and system levels. As such, design thinking can be used as a framework for human-centred innovation (Koh et al. 2015).

In the Graduate Certificate course, participating PNG teachers were introduced to the concepts and processes of design thinking. In this chapter, four teachers share narratives about learning experiences that they designed to suit their own contexts. We adopt the lens of learning about and through design (English et al. 2020) to examine how PNG teachers' applied design thinking in their teaching practice.

TEACHER NARRATIVE 7.1: Douglas Hopkos taught Grade 12 students at the Bishop Leo Arkfeld Secondary School in the East Sepik Province. He designed a learning experience for students to learn Physics concepts through designing an Evaporating Environment.

TEACHER NARRATIVE 7.2: Lonah Tombele guided activities for Grade 7, 10, 11, and 12 students at the Kilakila Secondary School. She designed an integrated STEM learning experience based around designing and establishing a School Conservation Park Project.

TEACHER NARRATIVE 7.3: Manasseh Kui Fena taught Grade 11 students at the Wawin National School of Excellence. He designed a learning experience for students to learn about design thinking through making a solar power bank project.

TEACHER NARRATIVE 7.4: Raphael Ulka worked with students and teachers at Gumine Secondary School. They learnt about and practised design thinking as a problem-solving approach to assure the school's water supply.

Before we present the teacher narratives, it is important to emphasise briefly that our focus on design thinking in STEM Education in this chapter is not intended to suggest that its use should exclude or replace other approaches to problem-solving or STEM thinking skills such as mathematical modelling, scientific method, computational thinking and systems thinking. Rather, design thinking is a complementary and important inclusion (Li et al. 2019). As emphasised by Kelley and Knowles' (2016) framework, each STEM field offers a unique but complementary contribution to STEM Education that relates to its unique perspective on problem-solving. The unique contribution of design thinking is its empathetic and human-centred approach to understanding problems and developing solutions.

The following teacher narratives are presented to illustrate how design thinking was understood by PNG teachers through the AAPNG STEM in Education project, and how they applied design thinking in their teaching practice and make clear links between discipline knowledge and its application to design thinking in the context of STEM Education.

7.2 Teacher Narrative 7.1

Douglas Hopkos

7.2.1 Design Thinking in STEM Through an Evaporating Environment Project

Teaching Physics in Papua New Guinea is very challenging. The number of students taking the subject is very small and the passing rate is declining each year. Without a proper science laboratory and equipment, lessons are mostly theoretical, posing a challenge for students to grasp practical concepts as hands-on demonstrations and laboratory experimentation are not done. The student-to-teacher Physics class ratio is small compared to other subject classes. However, students' participation in solving theoretical problems consumes more lesson time because more explanation is required at a slow pace for students to understand concepts before moving on to the next topic or unit. This slows down the content coverage and does not allow room for more practice in problem-solving. The net effect is seen in many unanswered questions in the national Physics exams and low passing rates.

7.2.1.1 Rationale

STEM integration is done here as a parallel activity alongside a Grade 12 unit on heat energy but captures concepts taught in previous units in Grades 11 and 12. The STEM integration activity I designed is based on creating an environment that manipulates factors affecting the rate of evaporation of water. The purpose is to let students explore and investigate various techniques and materials available in the environment to manipulate the rate of evaporation of water.

It is anticipated that, through learning and improving the evaporating environment throughout the unit, the students will learn by applying the unit concepts. Table 7.2 summarises the problem statement, the connections to the curriculum and global goals, the pedagogy I chose, and a brief reflection on the outcomes of the project.

7.2.1.2 Implementation

This implementation of STEM integration, while promoting design thinking, makes deliberate iterative use of the 5E instructional model (Bybee 2010). The steps of the 5E model, as applied in this project, are: Engage, Explore, Extend and Explain, and Evaluate.

Engage: The students were engaged in the concepts of the heat energy unit in Grade 12 Physics through observing and relating to common experiences of water evaporating from substances and the environment. Students could relate to experiences when they had to wash their uniforms and dry them after school using various drying techniques. This created discussion and some students started trying to explain the processes using scientific principles. Students confidently identified the process as evaporation.

Explore: Students were allocated to groups. They were then challenged to explore the factors that affect the rate of evaporation through simple investigations with common household utensils and other useful and common waste materials that could be found around them. The students worked in groups and contributed their knowledge, skills, time, energy, and internet accessibility and connectivity resources. After the students identified the factors, they went on to produce both written and simplified graphical explanations for each factor that they identified which affected evaporation.

Extend and Explain: The next stage was where the students extended their ideas on evaporation to apply them to designing and creating a physical environment where the rate of evaporation of water increased. The aim of the project was to investigate, improve and create optimum conditions to make water evaporate quickly. This was done by creating designs, discussing in groups, looking for available materials, constructing and evaluating. The students repeated the iterative cycle throughout the unit until they were satisfied that they had reached their optimum conditions for evaporation.

Evaluate: The final stage was where the students described the operation of their optimum drying environment and suggested ways for improvement. Students described the shape and structure of the evaporation environments they created and what features of their design or product prototype affected the factors determining rate of evaporation. By doing this, students demonstrated their ability in transferring learnt knowledge and skills across to real-life applications. A significant application of the factors affecting the rate of evaporation is in the cocoa and vanilla cash crop industry drying processes.

7.2.1.3 Outcomes

The outcome of the project was evident in the individual models that the groups produced. Each was supported by deep learning and thorough understanding of the Physics principles. The following shows student solutions (Groups A-D) and provides details of their processes.

Group A

Group A used waste cardboard, wood and aluminium foil (Fig. 7.1). They used a metal tray to evaporate water under direct sunlight and two mirrors to reflect light underneath the base. The open top allowed exposure to flowing air. They utilised the concept of metals being efficient conductors of heat and moving air helping water to evaporate fast.

Table 7.2 Narrative 7.1 project summary

Problem	Connections		Pedagogy	Reflection
Using natural elements to efficiently dry up/ evaporate liquids	*Curriculum Connections*		5E Inquiry-Based Model (Bybee 2010)	Application of principles of heat energy to physical experiments
	Grade 11 and 12 Physics			
	SDGs (UN, n.d.)			
	4	Quality Education		
	7	Affordable and Clean Energy		
	12	Responsible Consumption and Production		

Fig. 7.1 Group A's evaporation environment

Group B

Group B used clear plastics from grocery shopping bags with cardboard (Fig. 7.2). They wrapped clear plastic over the top of a box with an air inlet and outlet on the sides. The trapped heated air increased the temperature quickly while the air holes allowed heated air to move out as cool air moves in to evaporate water. They used an empty can and transferred measurements from a beaker to a new container to measure changes in the volume of water.

Group C

Group C used cardboard boxes holding a calibrated container of water and river pebbles to capture heat (Fig. 7.3). They placed a container of water in the open air and sunlight. They placed dense river stones to capture the heat from the sun and warm the container.

Group D

Group D used cardboard boxes for mounting the water container and aluminium heat reflectors (Fig. 7.4). They created a self-contained light reflector with aluminium reflecting light to the base of the container combined with direct sunlight from the top.

7.2.1.4 Reflection

Learning Physics through a practical project applied to a Grade 12 unit on heat energy has helped my students to better understand the concepts taught. Students applied their knowledge on factors affecting rates of evaporation to the local environment by trying different materials beyond what was anticipated by the teacher. Understanding that heat conductivity and temperature can change the behaviour of materials helped students to work out which materials to use in

Fig. 7.2 Group B's evaporation environment

Fig. 7.3 Group C's evaporation environment

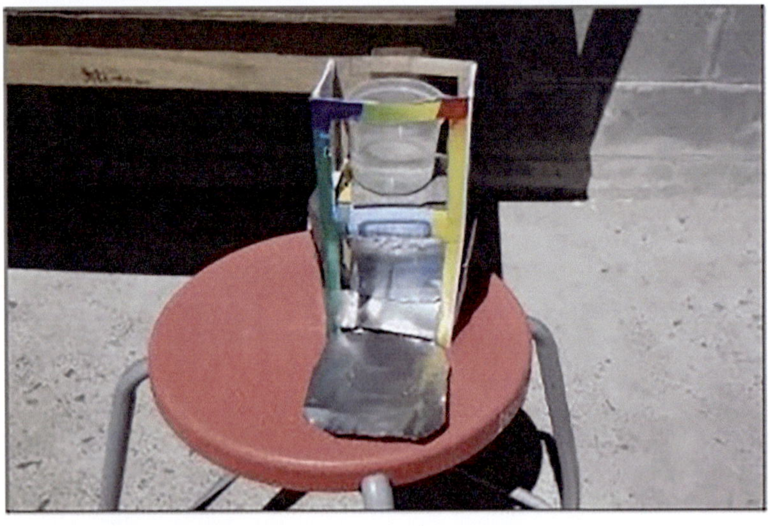

Fig. 7.4 Group D's evaporation environment

Fig. 7.5 SDGs met through a school conservation park project

the project. Students explored commonly available materials such as empty soft drink cans, cardboard boxes, stones of various densities and colours, plastic shopping bags and aluminium foil and other materials to create their optimum environment. Students realised that these objects can be assembled in many ways while still succeeding in manipulating factors affecting the rate of evaporation. By measuring the rate of evaporation of water each day, the students were able to modify their designs creating a lot of excitement and enthusiasm in students unlike before.

The successful outcomes for my students in this activity were partly due to their willingness to collaborate in teams. This meant that they learnt from one another (Wanzek et al. 2014), engaged more meaningfully with each other, and took ownership of the knowledge that they generated together (Rosicka 2016; Sirinterlikci et al. 2009). Collaborative learning is a productive way to approach Design Thinking in STEM, rather than the usual fragmented ideas found in outcome-based education targeted at individuals.

The project also contributed to the achievement of SDGs (Fig. 7.5). These are SDG 4 [Quality Education], SDG 7 [Affordable and Clean Energy] and SDG 12 [Responsible Consumption and Production].

This STEM activity enhanced student motivation in learning Physics. This was shown in the way students collected and assembled materials for the project. It also made students provide explanations based on their understanding of concepts onto the type of modifications they were making on the evaporating environment to reach an optimum condition. They applied their Design Thinking and tested their theories through a practical test, all of which consolidated their learning.

7.3 Teacher Narrative 7.2

Lonah Tombele

7.3.1 Design Thinking in STEM Through a School Conservation Park Project

At Kilakila Secondary School, we undertake STEM projects that integrate the Science and Technology curricula (PNGDOE 2023). The school is in the heart of squatter settlements where we experience noise pollution and overcrowded shanty homes. Despite this the school has, over the last ten years, improved its standard in terms of infrastructure development, including a mini-grandstand and new classroom facilities. As a result of these developments, the natural species of plants and animals that were around are no longer here. The noise of machines, trucks and the surrounding settlements have scared them away and disturbed the ecosystem. Pollution levels have also gone up. Changes such as these can directly contribute to climate change which is a global concern. As a school, we need to promote environmental conservation and reduce the causes of climate change in the best ways we can to help sustain our region's natural beauty and its ecosystem. I opted to apply a social constructivist model where learning is socially situated, and students work together to enhance their own learning and to complete a shared project (McKinley 2015).

7.3.1.1 Rationale

Creating a green space by developing a mini conservation park was our goal. My aims, through an integrated STEM project involving students in Grades 7, 10, 11 and 12, were to:

1. plant different species of forest trees so that the animals and plants we lost through development and other drastic changes can come back to the environment they inhabited years ago;
2. assess students in all disciplines using the component of content applicable to their discipline. Different rubrics were collaboratively developed to suit each subject and grade assessment (see also Narrative 9.1).

Table 7.3 summarises the problem statement, connections to the curriculum and SDGs, the pedagogy applied and a reflection on the project.

7.3.1.2 Implementation

There were several processes involved in this school-based conservation project undertaken by different year levels. The project involved four different grades, 11 classes and four subjects. The project's processes were: (1) site selection and measurement; (2) approval and funds for ploughing; (3) modelling; (4) sourcing our trees; and (5) planting.

Site selection and measurement: The selected site of the project was the school's old botanical garden. Late in 2011, the space was converted into the school's agriculture farm. With a reduction in Technology classes at the school, the farm was left fallow. Over the years, the teachers have made small gardens there and most times it is bushy with shrubs, weeds and long grass. Using wool we bought from the shop, the Grade 11 Science boys measured the dimensions. Their measurements were later confirmed by re-measuring with a tape measure. The area has a length of 245 m and a width of 30 m (an area of 7350 m^2). It is located right in front of the teachers' residential area and the track that leads to the teachers' houses marks its border. We first developed the right side of the area because there some teachers were yet to harvest their yams.

Approval and funds for ploughing: After selecting and measuring the site, I had a brief discussion with the school principal. I argued that the now-unused land would be ideal for trees because, in the late 1990s, it had been a botanical garden. Therefore, it was my land of choice for our current project to create a green space in the school which we would call a mini-School Conservation Park. The school principal agreed and approved funds of K350.00 (85 USD$) to hire a machine to begin ploughing the next day.

Modelling: Students in Grade 11 Geography and Grades 11 and 12 Science were asked to create potential models of our mini-School Conservation Park. Not only would the models be assessed, but there would also be a competition with the winning entry put into practice. The students worked in groups and were only permitted to use recycled and waste materials in the making of their models. The models were exhibited at a school assembly.

We invited people from the Environment and Conservation office, PNG Nature Park and the media team from TVWan (a digital free-to-view TV network with coverage across PNG). Judges went around assessing each group while the students actively explained how they constructed it and what materials they used to make the model. The whole school provided the audience for the exhibition. All were upcycled models because they were created from waste materials from around the school. The "winning" mini-School Conservation Park model was displayed at the APEC Haus, Ela Beach, Port Moresby on World Environment Day (Figs. 7.6 and 7.7).

Sourcing our trees: After selecting the model, we imagined the type of environment it would be. During this process, we were called by the PNG Nature Park who were willing to donate 1000 tropical tree species to the park. However, our land mass could not cater for that number at the moment, so we got them in instalments. We only received 223 seedlings. After we cleared the other side, we would be able to claim the remainder.

Table 7.3 Narrative 7.2 project summary

<table>
<tr><th>Problem</th><th colspan="2">Connections</th><th>Pedagogy</th><th>Reflection</th></tr>
<tr><td rowspan="7">Environmental degradation due to human development</td><td colspan="2">Curriculum Connections</td><td rowspan="7">Social constructivism (working in groups and learning with others)</td><td rowspan="7">Students were able to engage in transdisciplinary learning to regenerate a green space as a mini-School Conservation Park</td></tr>
<tr><td colspan="2">Grade 11 and 12 Science and Geography</td></tr>
<tr><td colspan="2">SDGs (UN, n.d.):</td></tr>
<tr><td>4</td><td>Quality Education</td></tr>
<tr><td>12</td><td>Responsible Consumption and Production</td></tr>
<tr><td>13</td><td>Climate Action</td></tr>
<tr><td>15</td><td>Life on Land</td></tr>
</table>

Fig. 7.6 Judging of student models of the Conservation Park

Planting: As part of a Grade 10 Agriculture assessment task, students marked out and planted our seedlings (see Narrative 9.1). They planted the trees randomly to create a mini forest rather than a mono-cropping forestry. The seedlings were spaced 1.5m apart in rows which were roughly 3m apart.

7.3.1.3 Outcomes

This project has three significant outcomes. First, meeting my first aim, this project has succeeded in establishing a mini-School Conservation Park in what was an abandoned garden. The regenerated green space is a tangible outcome for the school to enjoy and make a positive environmental outcome. Because of their hands-on involvement in different subjects or as active observers, students feel a sense of pride and ownership in the park. The work of planting and maintenance continues.

Second, responding to the demands of the PNG Standards Based Curriculum (PNG Education

Fig. 7.7 The winning model selected for implementation

News 2021), this project has also demonstrated the possibilities of integrating STEM subjects through a design thinking approach. Students and teachers across different subjects and year levels were actively involved in the planning and evaluation of the project. The school administration scheduled school-based in-service workshops for teachers and supported students who exhibited their models.

Finally, the project, meeting my second aim, allowed me to create a community of practice within my school of teachers collaborating on assessing students on similar topics across subjects and grades (see Narrative 9.1). Creating the rubrics was initially time-consuming, but this investment of time was rewarded in their multiple uses and the consistency achieved.

The problems we encountered (and overcame) in achieving these outcomes included: maintaining consistency and parity in the pace of teaching across different grades; delays in scheduling some components; and finding the funds for the project. The costs were not budgeted in the School SLIP (Student Learning Improvement Plan) programme. Fortunately, we have several collaborative partners. As noted, the PNG Nature Park, promised 1000 seedlings. These cost K10 (2.5 USD$) each in the Port Moresby markets and would have been beyond our means.

7.3.1.4 Reflection

My reflections on this project are wide-ranging. I am pleased with the school's achievement in regenerating a green space and improving our learning environment. That this has been done through a collaborative effort is gratifying. Everyone in the school is proud of the park we created. Together, we have enacted the principles of the PNG Standards Based Curriculum, which I value as a body of knowledge and set of competencies that can form the basis for a quality education.

I have seen how integrating STEM across the curriculum needs teachers to collaborate to achieve shared goals. I now know that custom rubrics can help students understand teachers' expectations in any assessment task. They make scoring easier and reduce arguments with students regarding their assessment. They also allow teachers to clarify the real purpose of the activities they set.

I wholeheartedly support the role of Design Thinking as a strategy in integrated STEM. Student learning has been amplified and many of the skills gained will have ongoing value to them. There is a broader value to this project as measured by the SDGs (Fig. 7.8). We have contributed towards achieving SDG 4 [Quality Education], SDG 12 [Responsible Consumption and Production], SDG 13 [Climate Action] and SDG 15 [Life on Land].

Finally, I remain concerned for those students who do not take STEM subjects or who are not involved in projects such as this but are interested in STEM career pathways. I wonder how we might equip them for further study or future careers. The STEM syllabus is offered in the six national Schools of Excellence in PNG but not in our secondary schools. I hope that one day it is taught more widely across the country and students have more equitable access to the type of experiences that our students have enjoyed in this project.

7.4 Teacher Narrative 7.3

Manasseh Kui Fena

7.4.1 Design Thinking in STEM Through a Solar Power Bank from Recycled Laptop Battery

Electricity consumption is increasing in PNG. It is imperative that PNG searches for new and reliable energy sources to meet our requirements while considering the economic costs and the potentially negative impact on the environment. As for other teachers in PNG (see Narratives Abel P. Mambo 8.3; Debbie Zaccias 6.2; Koni Wamzii N'Drihin 6.5), I wanted to teach my Grade 11 Applied Science students about renewable and sustainable energy. One such energy source is harnessing the power of the sun via solar panels (Ismangil and Susanto 2019).

I chose to design a hands-on learning experience where students would produce a power bank with repurposed recycled laptop batteries (Stewart 2021). I believe that this approach, using a Design Thinking strategy, would enhance students' knowledge, give them awareness of global issues and the opportunity to reuse waste materials. Above all, I wanted my students to be challenged by a real-world problem and to be empowered by developing a working solution. For example, our school in Lae has an unreliable electricity supply and we experience frequent power outages. When there is a power blackout, students are forced to study by cellphone torches or to use candles which pose a fire risk. Many of our students come from rural to remote areas where there is no electricity. Taking a power bank home will have an immediate impact on students and their families. Table 7.4 summarises the problem statement, the connections to the curriculum and sustainable development goals, as well as the pedagogy I employed and a brief reflection on the project.

7.4.1.1 Rationale

The Applied Science syllabus (PNGDOE 2008b) includes the study of Alternative Renewable Energy Sources. I decided, in covering Unit 11.3 Energy Around Us, that I would ask my students to develop a solar power bank. This would provide a practical application of the theory in the unit and address some critical issues for our school and, more broadly, for the country.

The first issue addressed is the need, as noted, for reliable, clean and affordable power generation in PNG. This is directly connected to the SDG 7 relating to Affordable and Clean Energy. It is also a priority of the PNG Vision 2050 (Independent State of Papua New Guinea 2010) and is included in the PNG Medium Term Development Strategies (Independent State of Papua New Guinea 2018). Emetere et al. (2018) investigated the effectiveness of solar power banks as "the cheapest way to resolve the lingering challenges of solar energy storage for rural dwellers in parts of Nigeria" (p. 17). In Australia, social enterprise and charity "Substation 33" has developed a "90 percent plug-and-play home solar system" (Stewart 2021, para.18) which has been exported to Indonesia.

The second issue addressed is the need for students to understand how building a solar power bank has an indirect link to achieving SDG 15 which relates to Life on Land. In this project,

Fig. 7.8 SDGS met through a school conservation park project

Table 7.4 Narrative 7.3 project summary

Problem	Connections		Pedagogy	Reflection
Respond to growing demand for electricity and current unreliable electrical supply at the school	*Curriculum Connections*		*d.School* model of Design Thinking (Hasso Plattner Institute of Design 2022)	Students were provided with an opportunity to plan and construct energy-producing models
	Grade 11 Applied Science, Chemistry, Physics, Advanced Maths, and Design & Technology			
	SDGs (UN, n.d.)			
	7	Affordable and Clean Energy		
	10	Reduced Inequalities		
	12	Responsible Consumption and Production		
	15	Life on Land		
	17	Partnerships for the Goals		

students will be recycling lithium-ion batteries from old computer monitors and television sets. This means that they will be recycling potentially toxic waste materials into a useful product instead of disposing of them in landfills. Reusing and recycling laptop lithium batteries to produce electricity for lighting keeps them out of the environment and stops them polluting the soil and contaminating groundwater. It also reduces the number of non-renewable resources used in producing new batteries.

7.4.1.2 Implementation

The first step in any integrated STEM activity is to identify the project's curriculum connections. This project is also connected with the PNG curriculum in Grade 11 in five subject areas:

- Advanced Mathematics: 11.1—Number and Applications (PNGDOE 2008a)
- Applied Science: Unit 11.3—Energy Around Us (PNGDOE 2008b)
- Chemistry: Unit 11.4—Energy and Reaction Rates (PNGDOE 2008c)
- Design & Technology: Computer Fundamentals. (PNGDOE 2008d)
- Physics: Unit 11.4—Work, Power and Energy; Unit 11.5 Electricity Principles (PNGDOE 2008e).

By undertaking real-world problems or projects students apply knowledge and skills from two or more disciplines and help to shape the learning experience (Vasquez et al. 2013). The design thinking approach I adopted was the d.School model of: Empathise, Define, Ideate,

Prototype, Test and Pitch (Hasso Plattner Institute of Design 2022).

Empathise: To begin, I introduced and established an emotional connection to the topic by talking about life without electricity in a rural community. Students were asked to reflect on their own experiences at home and at school. How might life be improved with reliable electricity? The student worksheet I prepared prompted:

> How do you feel about doing your study under a candlelight or phone torch? What are the bad effects of doing that? Think about your parents and the rest of the family back home. How do you feel about them living without any electricity? What do you think you can do to solve the problem given your context?

Define: I asked my students to define the problem. I then explained and demonstrated how a solar power bank works. They understood that solar energy will be absorbed by the photovoltaic cells in the solar panel and converted into chemical energy which is stored in the assembly of lithium batteries. This chemical energy is then converted to light and heat energy and chemical energy again when phones are charged. We discussed the broader impact this invention has on sustainability (through recycling). The student worksheet posed some critical questions:

1. What is a solar power bank?
2. What STEM concepts are involved in this project?
3. What are the resources needed in this project?

Ideate: I guided my students through a brainstorming session to consider potential solutions. Their thoughts ran wild but eventually settled on a shared problem statement:

> How might we design an innovation in our context that will solve the energy problem in providing light within the next two months?

The student worksheet, which provided formative assessment, instructed students to:

> Investigate and do research on the project.
>
> 1. Where do you get the information you need for the project?
> 2. What kind of information will come from your research and how do you use it?
> 3. How can you utilise the available resources to maximize output?

Prototype: My students formed into pairs. I issued the materials (such as used laptop batteries and circuit boards) they needed to build their prototype designs. The student worksheet advised:

> Make the preferred solution. In making or creating the prototype, what problems did you encounter and how did you address them?

Test: Students were allowed to independently test their designs. They were challenged with prompts such as:

- If it works, how can it be done better or what else needs to be improved?
- If it doesn't work, what went wrong and how can it be fixed?

They were encouraged to be ready to return to previous ideation step based on the results of their testing. The student worksheet advised:

> Implement the solution you have designed and developed. Sometimes, if the outcome does not turn out right, previous steps have to be revisited again. If it works as expected, recommendations can be made. However, several tests can be done for a good positive result. How or what else can it be done to improve efficiency and effectiveness?

Pitch: Once assembled and tested, students were asked to "pitch" their designs to the principal, other interested teachers and students. This, along with a digital portfolio and slide presentation, made up the summative assessment for this activity.

7.4.1.3 Outcomes

Through this project, students used design thinking to examine, plan and construct energy-producing models, and researched and investigated alternative sources of energy production. Design thinking provided opportunities for them to acquire new skills and knowledge fulfilling the expected outcomes detailed in the PNG syllabus. Real-world learning experiences consolidated students' learning in the classroom allowing them to demonstrate an understanding of fundamental principles and models of science (PNG Department of Education 2023).

This approach also allowed teachers to reflect on their professional practice. It helped teachers to empathise with students' experiences and to identify challenges and respond by designing student- and context-specific practices. Design thinking makes teachers active participants and stakeholders in new practices (Blundell 2022).

7.4.1.4 Reflection

This project is significant. The national government's agenda of the rural electrification programme has been ineffective, and people continue to live in dire need of electrical energy. Solar power banks will give light to households in rural villages and give certainty to those who are connected to the main grid but who endure frequent power outages. It will light up their homes, especially at night, and charge their cell phones. Stewart (2021) reported that the experience of using solar power banks in Indonesia has meant that when people "have lighting overnight, they study more, they socialise more and they do more work that they can be paid for, things like making woven mats … it increases their well-being, but it also increases their source of income" (paras. 47–48).

The project will also contribute to the achievement of SDGs (Fig. 7.9). These are SDG 4 [Quality Education], SDG 7 [Affordable and Clean Energy], SDG 10 [Reduced Inequalities], SDG 12 [Responsible Consumption and Production], SDG 15 [Life on Land] and SDG 17 [Partnerships for the Goals].

Solar power banks provide a clean sustainable source of renewable energy. In PNG where, as noted, there are frequent power outages, there is a need to develop a cheaper, sustainable, and environmentally safe source of household energy. The power banks will reuse laptop batteries to provide light and to charge mobile phones.

The stumbling block to this learning experience is access to specialised resources. We needed: power bank kits—to provide the energy source; Solar panels—to convert energy; and, Cables—to connect components. In addition, the power banks need circuit boards and casings which are made in China but shipped from Australia. This means we are dependent on when cargoes arrive in Australia and are dispatched to PNG. These components have a cost and so funds will need to be sought for a bulk purchase from some funding organisation. Funds will also be needed to distribute to schools in the surrounding community, especially those who do not have access to electricity.

Another resourcing challenge will be mobilising laptop batteries. Papua New Guinea generally does not recycle e-waste and batteries end up in landfill. A solution would be to post on social media platforms promoting the idea of collecting used laptop batteries. So, instead of people throwing the damaged laptop batteries away, they can give them to PNG schools for reuse and repurposing for the solar power bank project.

All teachers embarking on integrated STEM projects like this should ask themselves the searching questions that I asked my students. These are:

- What were the ideas or steps that worked well in this task?
- What were the challenges you encountered?
- Do you have any further clarification questions?
- How can the project be done better the next time around?
- What were some of the areas that require improvements?
- Were resources readily available and easily accessed?
- Can the students be able to do similar projects in their community?
- What would be some of the limiting factors to successfully delivering any similar project in their community?
- Has every student grasped the STEM concepts encapsulated in the project?

I have learnt that PNG teachers are resilient. When an idea does not work as anticipated, students are encouraged to redesign and improvise with available resources keeping the whole

Fig. 7.9 SDGs met through a solar power bank project

idea of producing a solar power bank alive. I believe that the students engaging in the activity have expanded their knowledge about sustainable renewable energy and deepened their thoughts on alternative energy sources and can be able to engage in projects that can provide sustainable cost-effective energies.

7.4.1.5 Postscript

Stemming from my original school-based project, the teachers initiated a SolarUp Wawin project last year. I was the fundraising chairman of the SolarUp Wawin team. The aim was to raise funds to purchase and install solar sets in all 30 staff houses. The estimated cost is K1.5 million (365, 000 USD$). The fundraising was still continuing when I left for my new appointment at Kopen National School of Excellence.

At my new location, I am writing a proposal to seek funding to light up my new school. We are heavily dependent on burning diesel fuel for electricity and would prefer to use sustainable renewable solar power. This new project will install a solar powered generator that will provide electricity to 30 staff houses, 22 classrooms, two computer laboratories, four science laboratories, administration building, seven student dormitories and 1.5 km of street lighting around the perimeter fencing. This is an ambitious project which will build on my previous experience at Wawin National School of Excellence.

7.5 Teacher Narrative 7.4

Raphael Ulka

7.5.1 Design Thinking in STEM to Solve Water Supply problem in Gumine Secondary School

This narrative recounts how a small group of teachers and students at Gumine Secondary School in the southern part of Simbu Province got together to develop a solution to the real problem of restricted access to water at our school. The plan was enacted, in part, through the support of the PNGAus Partnership Secondary School (PASS) initiative programme (Commonwealth of Australia 2021).

Gumine Secondary, like many schools in PNG, has infrastructure problems. Our water, drawn from a nearby mountain and shared with the local community, is usually stored in reservoir tanks with controlled access. Water is only supplied to the school buildings and houses in the morning from 6 to 7am. In the afternoon, water supply is usually provided from 4pm or 5pm and cut off at 7pm or 8pm. Water is not accessible at any other time during the day. This is a pressing problem, and, at Gumine, we cannot wait for others to step in and sort it out for us.

Fortunately, Gumine Secondary was one of the 12 PNG schools included in the PNGAus Partnership Secondary School (PASS) initiative programme (Commonwealth of Australia 2021). The goal of PASS is to connect "Papua New Guinean and Australian students, teachers and school leaders to better prepare our future generations to thrive in a technology-driven and globally interconnected twenty-first century" (Queensland Government 2019, p.6). Gumine has been partnered with Bundaberg State High School in regional Queensland (Australia) (PASS 2023). In 2019, there were reciprocal visits by staff and students from both schools. The PASS project which provided direct support to our water supply problem is intended to:

> Improve sanitation and hygiene at Gumine Secondary School and strengthen intercultural understanding and relationships between students and teachers from both schools, through fundraising for a water tank and plumbing equipment, development of educational resources and lessons to improve hygiene awareness, as well as sharing of resources and demonstrations around PNG traditional arts and practices supported by regular communications and reflections on pedagogy and classroom practices.
>
> (PASS 2023, para. 11)

7.5.1.1 Rationale

I chose to adopt a Design Thinking problem-solving strategy because I believed it can provide an effective approach to help us to find a reliable home-grown solution. Design Thinking has been described as a "type of creative problem solving". Meg Dryer explained that:

> It's a way of thinking and making that keeps the user at the centre of everything … It's a human-centered approach … [it] is really deeply understanding people and their needs, synthesising down what problems you're going to solve, ideating around what concepts to design and then prototyping and testing over and over again while getting real user feedback.
>
> (Adobe 2025, para.1)

Another reason to adopt a Design Thinking strategy was because I wanted other teachers to see how it could enhance their students' skills and knowledge and how it could provide them with a structure for planning meaningful learning experiences in their own disciplines or in integrated STEM activities. Design Thinking allows students to be involved and to transfer what they learn in other disciplines to solve real-life problems in the school.

In Table 7.5, the problem statement is presented. Connections to the curriculum and SDGs are noted along with the chosen pedagogy for implementation and a reflection on the outcomes of the project.

7.5.1.2 Implementation

I began by gathering a team of students and teachers, all involved in the PASS programme, who were interested in working on this project. I explained the d.school model of the design thinking we were going to use. Its phases are: Empathise, Define, Ideate, Prototype and Test (Hasso Plattner Institute of Design 2022).

Empathise: At our first meeting, we identified who will benefit from an improved water supply. We interviewed them. The team did further research into some of the previous attempts made to solve this problem at the school. This gave us more information about the background and what we should potentially avoid.

Define: We reviewed the insights we had collected from the interviews and our research. I then asked the team to devise a "How Might We" question to help define the problem. The question that the students came up with was:

> 'How Might We' find a better way to make sure water is supplied to houses in the school for the whole day, every day, without disruptions and control?

Ideate: The "How Might We" question helped the team to generate possible solutions. They understood the users' needs (from the interviews) which enabled them to think creatively and generate ideas that offered a viable solution. The ideas were written on post-it notes and placed on a chart. We then went further and organised them into categories to make more meaning and get closer to identifying a possible solution. The ideas were put into different possible categories:

1. cheap and easy to implement
2. cheap but difficult to implement

Table 7.5 Narrative 7.4 project summary

Problem	Connections		Pedagogy	Reflection
Limited infrastructure means restricted access to water stored in reservoirs for only a few hours each day	*Partnership Connections*		d.school model of Design Thinking (Hasso Plattner Institute of Design 2022)	While the teacher was posted to another school, reports indicate that the project to improve infrastructure to increase access to water at the school was successful
	Responding to the PASS agenda to improve STEM learning (Commonwealth of Australia 2021)			
	SDGs (UN, n.d.)			
	6	Clean Water and Sanitation		
	9	Industry, Innovation and Infrastructure		
	10	Reduced Inequalities		
	17	Partnerships for the Goals		

3. easy but expensive to implement
4. expensive and difficult to implement.

Only ideas from the "cheap and easy" category were selected as the most reliable solution that was affordable to implement. We agreed that we should:

> Redo the water piping system of the school and install 9000 litre commercially available water tanks in various locations for easy distributions of water.

This idea was put to further discussion. Members of the team were thinking about how it could be done, who it would benefit, when it could be done and how it would be maintained. This provided the opportunity to give further thought to the idea, and to fine-tune it, until it became very clear.

Prototype: I then asked the team to use available resources to build a prototype. They used empty water containers as tanks, IV fluid tubes as pipes, sticky tape, nails, board, glue and paper clips. They wrapped off-cut timber with paper to indicate buildings namely the dormitories, the teachers' houses and the school rooms.

They collaboratively discussed and shared ideas on how they could build the prototype. They began by drawing a map of the school. This involved scales and dimensions, measurements, angles and surveying (from Mathematics). They also estimated the cost, including a budget for plumbing materials and labour. The team used their knowledge of science to discuss water pressure, gravitational force, purification of the water with chemicals and testing for water acidity. They used their knowledge in Design and Technology to build the prototype using tools and taking safety protocols into account.

Test: After the completion of the prototype, the team gathered to test the model. Water was poured into the two larger containers as reservoir tanks, and it flowed down through the pipes. Along the way they identified leakages and sections of the pipes where water was not flowing. These problems were rectified before we tested the model again. This trial-and-error process is an important part of Design Thinking. This time, water was able to flow freely through the pipes. However, the distribution of the water to the tanks was uneven; we readjusted the position of the tanks with the pipe linings. It then worked perfectly well as anticipated. The model was tested and finalised. It was then ready for use in presentations. We created a video to further explain our solution and the processes we followed.

Pitch: The next task was to make the project a reality. We approached the School Board to endorse our plan. Funding the project and engaging specialised personnel and labour was a challenge. Another challenge was to find a possible time to undertake the project. The PASS initiative came to our aid. Bundaberg State High

School bought two tanks for us. The Partnership page on the PASS website explained that:

> It was a momentous occasion when two large water tanks were delivered to Gumine Secondary School and installed in November 2020. Great teamwork and collaboration enabled this project to be completed with the Gumine team working on local logistics (capacity, supplier, location, cost, etc.) and the Bundaberg team raising awareness about and funds for the project. The two teams have held a couple of virtual meetings... during the year to discuss this and other projects.
>
> (PASS 2023, para.12)

7.5.1.3 Reflection

The teachers and students who participated in the project were able to embrace the Design Thinking problem solving skills in reaching an effective, affordable and easy to implement solution. They can use and apply the idea wherever they are in the future. It is a reliable strategy that can be used in any school setting, and therefore, I am planning to provide workshops to schools in the province so they can also use Design Thinking in their pedagogical approaches and solve real-life problems.

Along the way, students were given the opportunity to collaborate, communicate, use critical thinking and find creative ways to address the situation. They were able to incorporate what they have learnt in Science, Technology and Design, and Mathematics. Through their awareness of the project, teachers were also able to realise the importance of how design thinking processes can help solve real-life problems. They were also able to see the power of working in teams both close to home and external to the school.

The project also contributed to achieving the SDGs (Fig. 7.10). Specifically, SDG 4 [Quality Education], SDG 6 [Clean Water and Sanitation], SDG 9 [Industry, Innovation and Infrastructure], SDG 10 [Reduced Inequalities] and SDG 17 [Partnerships for the Goals] were addressed by this project.

Unfortunately, I was posted to another school and did not stay to facilitate or witness the implementation of the project. However, from reports I have received, the new tanks and pipes were installed in the school area. Now water distribution to the school is improved and the students and teachers are enjoying continuous access.

7.6 Review of Design Thinking in STEM in PNG

Christopher N. Blundell and Andy Yeh

Collectively, these narratives (7.1–7.4) illustrate how teachers in the course developed innovative approaches to teaching and learning through curriculum and pedagogy design. Their work contributes to a growing body of literature about professional learning partnerships between teachers and universities (Aksela 2019; Andino 2017; Kelly et al. 2019; Koh et al. 2015), the role of design thinking in curriculum and pedagogy innovation generally (Henriksen et al. 2017) and in STEM Education specifically (Kim 2021). Although each project is distinctive, they share a common focus on teachers designing opportunities to activate student engagement in authentic, contextually relevant projects that draw upon and develop STEM knowledge and skills. The teachers designed these projects as solutions to problems of practice.

7.6.1 Teachers Designing Solutions to Problems of Practice

Although design-based learning and professional learning literature does focus on STEM curriculum and pedagogy, STEM teachers independently designing classroom practices are reported less frequently (Kim 2021). The literature more frequently reports on STEM teachers engaging as co-designers with researchers or as participants learning how to implement design thinking from experts (Blundell 2022; Kim 2021). These four narratives provide an important contribution with each of the four teachers independently engaged in self-directed design activities. Consequently, there are variations in approach to problem definition and solution development

Fig. 7.10 SDGs met through a water supply project

phases. Indeed, the degree to which the narratives overtly focus on design thinking also varies.

Each narrative provides an insight to the teacher's designed solution to a problem of practice. The physics project-based learning strategy, designed by Douglas Hopkos and described in Narrative 7.1, is a response to a specific problem: students learning concepts experimentally and experientially. The other three narratives focus on enacting new opportunities for STEM Education: Lonah Tombele's conservation project (Narrative 7.2) and Manasseh Kui Fena's solar power bank project (Narrative 7.3) sought to identify ways to enact integrated/transdisciplinary STEM; Raphael Ulka's water supply problem project (Narrative 7.4) sought to engage teachers and students in design thinking in STEM. Although these are not overt problems, wanting to improve learning and teaching is a problem of practice. Collectively, those problems of practice relate to curriculum and pedagogy design in STEM Education.

In the problem definition phase (see Table 7.6), the strategies for empathising and for defining the problem vary. Both Lonah and Raphael sought to establish an empathetic position by interviewing people about their experiences and used this information to define the problem. Douglas and Manasseh relied on their personal experiences and understanding to identify literature-supported practices (project-based learning and design thinking pedagogies for integrated STEM, respectively) that they deemed relevant in their contexts. Only Raphael developed a problem statement. Henriksen et al. (2017) argued that the empathising phase is important for understanding all students' and teachers' experiences and for challenging assumptions because it allows the designer to "view their problem from multiple perspectives … [which leads to] deep understanding of the context and a vision for moving forward" (p.150). Kim's (2021) review of STEM teachers' design work identifies that the problem definition phase is less frequently reported than the solution phase, suggesting that the problem definition phase may be more tacit in nature with teachers relying on their deep knowledge of context. This would appear to be the case with these narratives.

There is also some variation in the solution development phase across the four narratives. All narratives focus on what is, in effect, the development of a prototype learning and teaching experience. As such they represent outcomes of the teachers' design work (Henriksen et al. 2017; Kim 2021). Raphael's project, which is an overt implementation of Hasso Plattner Institute of Design's (2022) design thinking model, has clear ideation, prototyping and testing phases with collaborative engagement of a range of participants. Lonah's narrative outlines how, due to the multiple year level, interdisciplinary nature

Table 7.6 Summary of Design Thinking narratives (problem definition phase)

Project	Problem definition phase	
	Empathise, discover	Define, interpret
Douglas Hopkos—evaporation projects in physics [Narrative 7.1]	Douglas identified need for more practical learning in Grade 12 Physics: framed in terms of students' experiences, problems of time, equipment limitations	
Lonah Tombele—conservation park project[1] [Narrative 7.2]	Lonah noted opportunity to initiate integrated STEM projects. Interviewed heads of department and reviewed curriculum for each subject	Lonah identified links between curriculum areas' content and pedagogies
Manasseh Kui Fena—solar power bank project [Narrative 7.3]	Manasseh noted opportunity to use design thinking for STEM projects	Manasseh identified links between curriculum content for STEM subjects
Raphael Ulka—water problem project [Narrative 7.4]	Students and teachers interviewed people to understand needs and researched previous attempts to solve problem	Students and teachers organised findings into themes and developed a problem statement

Note Information also drawn from Lonah Tombele's Teacher Leadership narrative (refer to Narrative 9.1)

of her conservation park project, colleagues were involved in resolving curriculum alignment and developing assessment rubrics. In their respective narratives, both Douglas and Manasseh describe their individual approaches to ideation, prototyping and testing (note: Manasseh's narrative is about a planned innovation). The variations in the solution development phase may relate to differences in the scope and nature of the reported innovations. Larger-scale projects that were directly relevant to multiple people relied on collegiate participation in ideation, prototyping and testing.

These narratives outline four teacher-designed solutions to problems of practice, while also providing insight to the open and varied nature of design thinking. As Kim (2021) noted, "the nature of design work is messy, and it can be a challenge for researchers to report on the process" (p. 151). Although models have been created to scaffold design thinking, such as d.school (Hasso Plattner Institute of Design 2022), design thinking is foremost a mindset and the process is non-linear, despite how it is often represented (Rauth et al., 2010). Although these models are representations of the work of experts, they may not fully represent the nuances of process nor the practice of designers (Razzouk & Shute, 2012). Consequently, it can take some time for teachers to understand the nature of design thinking (Retna, 2016). As such, these narratives represent initial but very important steps to the use of design thinking in STEM Education.

7.6.2 Students Learning STEM Through Design

The teachers' new approaches created opportunities for students to learn STEM *through* design. According to English et al. (2020), learning STEM through design focuses on developing disciplinary concepts while generating design solutions for real-world problems. Douglas' evaporation environment project for learning physics and Lonah's mini-School Conservation Park project for learning integrated science and technology illustrated this, learning STEM through a design approach. Douglas designed the evaporation environment learning experience with a clear purpose in mind, which was to develop students' understanding in the physics of heat energy and how it can be captured and dissipated to affect water evaporation rate. Students engaged in designing evaporation environments to conduct scientific investigations. Lonah's design of learning experience was focussed on the ecological and climate sciences and the applications of technologies (e.g.

measurement tools, machines for ploughing and digging). Through engaging in the design of a mini-School Conservation Park, students made sense of how STEM concepts are connected in solving complex real-world problems. Another feature of learning STEM through design can be seen from what and how the teachers assessed students' learning. Both Douglas' and Lonah's design of assessment tasks strongly emphasised the relevant STEM concepts. The project-based learning and design tasks/challenges became the means to engage in and to motivate learning. Design thinking was used for the purpose of learning about STEM concepts.

The narratives also illustrate three aspects of learning through design: authentic and contextual learning, design thinking as a process, and the place of prototyping. Firstly, each project employed an authentic, contextualised and relevant reason for students to engage in STEM learning. This is an important tenet for design thinking activities (Hasso Plattner Institute of Design 2022; IDEO 2023) and design thinking pedagogies (English et al. 2020; Koh et al. 2015). Secondly, Manasseh's and Raphael's projects directly involved students in design thinking to learn STEM through design. As English et al. (2020) noted, this creates opportunities for students to understand the interconnected nature of STEM knowledge. And finally, three of the four projects make purposeful use of prototyping. Raphael's project included opportunities for students to develop prototypes for water supply, which they then tested and refined. Although the projects described by Lonah and Douglas are not overtly framed as design thinking activities, both required students to create prototypes to test ideas. For Lonah's project, the prototypes were used to gather feedback on students' proposals for the park, and for Douglas's project, students created prototypes to test factors relating to evaporation. As such, students had first-hand experience creating and testing prototypes, which is an important aspect of design thinking (Razzouk & Shute 2012). Collectively, these narratives illustrate ways to incorporate aspects of design in project-based learning.

7.6.3 Students Learning about Design as an Approach to Solve STEM-Related Problems

Learning about design in integrated STEM project focuses on the understanding of design processes and applications of design thinking as a problem-solving approach (English et al. 2020). Teachers who design learning experiences in this form may include explicit teaching on the stages of design thinking first and then guide students to carry out these stages in solving a STEM-based problem. Manasseh's solar power bank project and Raphael's water problem project are typical examples of students learning about design. In both learning experiences, Manasseh and Raphael taught explicitly the five stages of design thinking (i.e. empathise, define, ideate, prototype, test) and students were instructed to use design thinking as the problem-solving approach to solve their respective energy and water problems. When "design" or "design thinking" is the topic of a learning experience, assessment usually involves assessing students' ability to carry out the design thinking process. This is reflected in Manasseh's narrative, in which he designed both formative and summative assessments on students' application of design thinking. Both narratives emphasised that design thinking is a key problem-solving approach, and they aim for teachers and students to possess and embrace design thinking skills for PNG's current and future STEM-related problems.

7.6.4 Implications and Opportunities for Future Practice

The narratives by Douglas, Lonah, and Raphael identify three challenges that have implications for future practice. The challenges relate to time, the need for support, and curriculum and assessment requirements. Each of the projects was conducted over an extended period of time to allow students opportunities to engage. All forms of inquiry-based learning require extended implementation periods. Similarly, the design thinking process requires participants (teachers and students) to devote time to understanding the problem and to designing the solution (English et al. 2020; Henriksen et al. 2017; Retna 2016). Consequently, when planning to implement design thinking as a pedagogy or as an approach to solving problems of practice, teachers need to allow sufficient time to engage in design thinking.

To enact these projects, the teachers required permission and support from school leaders and local authorities. Focussing on local problems made the learning experiences authentic and relevant to students, however, they also generated challenges that the teachers had to overcome by working with others. Teachers engaging with challenges is a key part of the design thinking process (Curtis et al. 2020). It can also require a degree of risk-taking when designing and implementing innovative approaches (Spoon et al. 2020). Consequently, when implementing similar projects, support from school leadership and relevant authorities is an important part of risk management (Retna 2016).

The integrated STEM projects (Lonah and Manasseh) required teachers to map how the projects aligned with the relevant curriculum and assessment requirements. By their nature, integrated STEM projects depend on teachers to identify curriculum alignment. McFadden and Roehrig (2017) noted that this requires a shift in focus from the design of specific tasks to the larger-scale mapping of curriculum, and subsequently, engaging with associated challenges. Lonah's and Manasseh's narratives illustrate this. Further, Lonah's narrative illustrates how this led to the prototyping phase: developing a common assessment task with different rubrics for each subject.

In addition to these challenges, the teachers' narratives also suggest three opportunities for future practice. The first relates to using design thinking to establish creative connections in STEM learning. The second relates to students engaging in projects that align with Sustainable Development Goals (SDGs). The third relates to teachers using design thinking in the realisation of SDG 4 [Quality Education].

Project-based learning, and similar inquiry-based pedagogies, often require the creative generation of ideas and potential solutions. Each of these narratives involves students being creative. Although only Manasseh and Raphael elaborate on how ideas were or could be generated, all four learning experiences are contingent on the generation of ideas. In addition to its potential as a pedagogy, design thinking provides strategies for participants to establish creative connections (Henriksen et al. 2017). The empathy phase helps to understand multiple perspectives that "that can unlock creative solutions" (Henriksen et al. 2017, p. 147) and the ideation phase, which involves divergent thinking, generates a range of possibilities. Similarly, design thinking is a way for teachers to generate creative solutions to pedagogical problems of practice (Kelly et al. 2019). Of course, accommodating the messiness of design thinking depends on respect and trust for participants (Henriksen et al. 2017; Kim 2021).

STEM has a role in the realisation of most SDGs, and consequently, the sustainable development goals and actions represent authentically important opportunities for STEM Education projects. All projects in this chapter directly align with the SDGs: Lonah's conservation park project addresses 'life on the land' (SDG 15); Manasseh's solar power bank project addresses 'affordable and clean energy' (SDG 7) and 'responsible consumption and production' (SDG 12); Raphael's water supply problem project addresses 'clean water and sanitation' (SDG 6). Given that most of the SDGs relate to people's lives, design thinking's human-centred approach to problem-solving is highly relevant to STEM Education projects that are focussed on sustainability.

Teachers using design thinking to generate solutions to problems of practice represent an opportunity to realise SDG 4 [Quality Education]. The narratives in this chapter highlight how seeking to improve the quality of education is a design process. Design thinking is a viable way for teachers as well as school and system leaders to engage in professional learning that is focussed on innovation (Yurkofsky et al. 2020). Its human-centred approach, in particular, the ways that empathising and ideation draw on experiences of people most central to the process (students and teachers), represents a responsive approach to improving the quality of education (Aksela 2019; Kelly et al. 2019; Koh et al. 2015).

7.7 Final Thoughts

More than simply a STEM curriculum topic, design thinking is an approach that teachers can use in their professional practice, and it is a way for students to learn STEM. This chapter illustrates how design can be used by teachers to develop solutions to problems of practice. It also illustrates how students can learn STEM through design, as well as develop design thinking skills as an approach to solving STEM-related problems. Through their inspirational work, Douglas Hopkos, Lonah Tombele, Manasseh Kui Fena and Raphael Ulka have identified, developed and implemented real STEM Education innovations in PNG. Their narratives provide insight to authentic and meaningful learning experiences and illustrate the importance of engaging in design thinking.

References

ACARA (Australian Curriculum, Assessment and Reporting Authority) (2023) Understand this learning area: technologies. https://v9.australiancurriculum.edu.au/teacher-resources/understand-this-learning-area/technologies#accordion-9c6911869c-item-68c55997bb

Adobe (2025) An introductory guide to design thinking. https://www.adobe.com/au/creativecloud/design/discover/design-thinking.html

Aksela M (2019) Towards student-centred solutions and pedagogical innovations in science education through co-design approach within design-based research. LUMAT: Int J Math, Sci Technol Educ 7(3):113–139. https://doi.org/10.31129/LUMAT.7.3.421

Andino CI (2017) Approaching bridges between university education and elementary school: design thinking for educators in El Salvador. Int J Des Educ 11(2):35–51. https://doi.org/10.18848/2325-128X/CGP

Billiar K, Hubelbank J, Oliva T, Camesano T (2014) Teaching STEM by design. Adv Eng Educ 4(1):9–12. https://www.scopus.com/inward/record.uri?eid=2-s2.0-84897437435&partnerID=40&md5=844975e1bdbb9004399cec8d20802f1f

Blundell CN (2022) A scoping review of design thinking in school-based teacher professional learning and development. Prof Dev Educ 1–16. https://doi.org/10.1080/19415257.2022.2132269

Bobbe T, Krzywinski J, Woelfel C (2016, May 16–19) A comparison of design process models from academic theory and professional practice [Conference Paper]. Design 2016. Dubrovnik, Croatia. https://www.researchgate.net/publication/313681084_A_Comparison_of_Design_Process_Models_from_Academic_Theory_and_Professional_Practice

Brown T, Katz B (2009) Change by design: how design thinking transforms organizations and inspires innovation. Harper Business

Bush S, Karp KS, Cox R, Cook KL, Albanese J, Karp M (2018) Design thinking framework: Shaping powerful mathematics. Mathematics Teaching in the Middle School 23(4):e1–e5. https://doi.org/10.5951/mathteacmiddscho.23.4.00e1

Bybee RW (2010) Advancing STEM education: A 2020 vision. Technol Eng Teach 13(6):30–35

Commonwealth of Australia (2021) Review of the PNGAus Partnership Secondary Schools initiative (PASS). https://www.dfat.gov.au/sites/default/files/review-png-australia-partnership-secondary-schools-initiative.docx

Curtis R, Cairns D, Bolyard J (2020) Understanding design, tolerating ambiguity, and developing middle school design-based lessons. ASEE Virtual Annual Conference. https://doi.org/10.18260/1-2--35420

Emetere ME, Osunlola OI, Adejumo AD, Dauda AO (2018) Design and construction of a domestic solar power bank in a convective environment. Open Fuels Energy Sci J 11(1):16–28. https://www.researchgate.net/publication/323736834_Design_and_Construction_of_A_Domestic_Solar_Power_Bank_in_a_Convective_Environment

English LD, Adams R, King D (2020) Design learning in STEM education. In: Johnson CC, Mohr-Schroeder MJ, Moore TJ, English LD (eds) Handbook of Research on STEM Education. Routledge, pp 76–86

Hasso Plattner Institute of Design (2022) d.school. Stanford University

Henriksen D, Richardson C, Mehta R (2017) Design thinking: a creative approach to educational problems of practice. Think Ski Creat 26:140–153. https://doi.org/10.1016/j.tsc.2017.10.001

Henriksen D, Gretter S, Richardson C (2020) Design thinking and the practicing teacher: addressing problems of practice in teacher education. Teach Educ 31(2):209–229. https://doi.org/10.1080/10476210.2018.1531841

IDEO (2023) Design thinking. https://www.ideou.com/pages/design-thinking

Independent State of Papua New Guinea (2010) Papua New Guinea National Strategic Plan 2010–2050 "2050 Vision". https://png-data.sprep.org/system/files/2011.png.vision.2050.pdf

Independent State of Papua New Guinea (2018) Medium Term Development Plan III (2018–2030) Vol.1. https://png-data.sprep.org/system/files/MTDP-III-Book-1_Final-Proof-Web-compressed.pdf

Ismangil A, Susanto HP (2019) Design of power bank mobile using solar panel-based microcontroller atmega 328. IOP Conference Series: Materials Science and Engineering 621(1):1–7. https://doi.org/10.1088/1757-899X/621/1/012008

Kelley TR, Knowles JG (2016) A conceptual framework for integrated STEM education. Int J STEM Educ 3(11). https://doi.org/10.1186/s40594-016-0046z

Kelly N, Wright N, Dawes L, Kerr J, Robertson A (2019) Co-design for curriculum planning: a model for professional development for high school teachers. Aust J Teach Educ 44(7):84–107. https://doi.org/10.14221/ajte.2019v44n7.6

Kim MS (2021) A systematic review of the design work of STEM teachers. Res Sci Technol Educ 39(2):131–155. https://doi.org/10.1080/02635143.2019.1682988

Koh JHL, Chai CS, Wong B, Hong H-Y (2015) Design thinking for education: Conceptions and applications in teaching and learning. Springer Singapore. https://doi.org/10.1007/978-981-287-444-3

Li Y, Schoenfeld AH, diSessa AA, Graesser AC, Benson LC, English L, Duschl RA (2019) Design and design thinking in STEM education. J STEM Educ Res 2:93–104. https://doi.org/10.1007/s41979-019-00020-z

McFadden JR, Roehrig GH (2017) Exploring teacher design team endeavors while creating an elementary-focused STEM-integrated curriculum. Int J STEM Educ 4(1):1–22. https://doi.org/10.1186/s40594-017-0084-1

McKinley J (2015) Critical argument and writer identity: social constructivism as a theoretical framework for EFL academic writing. Crit Inq Lang Stud 12(3):184–207. https://doi.org/10.7275/0qpc-ws45

PASS (PNGAus Partnership Secondary School) (2023) School partnerships. https://www.pass.edu.au/about-pass/partnerships

PNGDOE (Papua New Guinea Department of Education) (2008a) Advanced mathematics upper secondary teacher guide. https://www.education.gov.pg/syllabus/secondarysyllabus/subjectssyllabus/Advanced%20Mathematics%20Grade%2011%20&12%20Syllabus.pdf

PNGDOE (Papua New Guinea Department of Education) (2008b) Applied Science upper secondary syllabus. https://educationpng.gov.pg/Teachers/secondary/upper/syllabus-upper-secondary-applied-science.pdf

PNGDOE (Papua New Guinea Department of Education) (2008c) Chemistry upper secondary teacher guide. https://png.oer4pacific.org/id/eprint/324/1/syllabus-upper-secondary-chemistry.pdf

PNGDOE (Papua New Guinea Department of Education) (2008d) Design and technology upper secondary syllabus. https://www.educationpng.gov.pg/Teachers/secondary/upper/teachers-guide-upper-secondary-design-technology.pdf

PNGDOE (Papua New Guinea Department of Education) (2008e) Physics upper secondary syllabus. https://www.education.gov.pg/syllabus/secondarysyllabus/subjectssyllabus/Physics%20Grade%2011%20&%2012%20Syllabus.pdf

PNG Education News (Papua New Guinea Education News) (2021, May 22) When and why was Standard Based Education or Curriculum introduced in PNG? https://edu.pngfacts.com/standard-based-education/when-and-why-was-standard-based-education-or-curriculum-introduced-in-png

PNGDOE (Papua New Guinea Department of Education) (2023) Teachers. https://educationpng.gov.pg/Teachers/index.html

Queensland Government (2019) Bundaberg State High School: Annual Report 2019. https://bundabergshs.eq.edu.au/supportandresources/formsanddocuments/annualreports/annual-report-2019.pdf

Rauth I, Köppen E, Jobst B, Meinel C (2010) Design thinking: An educational model towards creative confidence. Paper presented at the 1st International Conference on Design Creativity (ICDC2010), Kobe, Japan. https://hpi.de/fileadmin/user_upload/fachgebiete/meinel/papers/Design_Thinking/2010_Rauth_ICDC.pdf

Razzouk R, Shute V (2012) What is design thinking and why is it important? Rev Educ Res 82(3):330–348. https://doi.org/10.3102/0034654312457429

Retna KS (2016) Thinking about "design thinking": a study of teacher experiences. Asia Pac J Educ 36(sup1):5–19. https://doi.org/10.1080/02188791.2015.1005049

Rosicka C (2016) Translating STEM education research into practice. Australian Council for Educational Research. https://research.acer.edu.au/professional_dev/10/

Sirinterlikci A, Zane L, Sirinterlikci AL (2009) Active learning through toy design and development. J Technol Stud 35(2):14–22. https://doi.org/10.21061/jots.v35i2.a.2

Spoon R, Rubenstein LD, Shively K, Stith K, Ascolani M, Potts ML (2020) Reconceptualizing professional learning within the gifted field: exploring the instruct to innovate model. J Educ Gift 43(3):193–226. https://doi.org/10.1177/0162353220933001

Stewart J (2021, March 6) Queensland entrepreneurs use old laptop batteries to help combat global energy crisis among poverty-stricken families. ABC News. https://www.abc.net.au/news/2021-03-06/qld-kingston-ewaste-global-energy-poverty-indonesia-powerwells/13206106

United Nations (UN). (n.d.) Sustainable development goals. https://www.un.org/sustainabledevelopment/sustainable-development-goals

Vasquez JA, Sneider C, Comer M (2013) STEM lesson essentials, Grades 3–8: Integrating Science, Technology, Engineering, and Mathematics. Heinemann

Wanzek J, Vaughn S, Kent SC, Swanson EA, Roberts G, Haynes M, Fall A-M, Stillman-Spisak SJ, Solis M (2014) The effects of team-based learning on Social Studies knowledge acquisition in high school. J Res Educ Effect 7(2):183–204. https://doi.org/10.1080/19345747.2013.836765

Yurkofsky MM, Peterson AJ, Mehta JD, Horwitz-Willis R, Frumin KM (2020) Research on continuous improvement: exploring the complexities of managing educational change. Rev Res Educ 44(1):403–433. https://doi.org/10.3102/0091732X20907363

Christopher N. Blundell is an Associate Professor at QUT. He is the Academic Lead for Learning and Teaching, and the Master of Education Course Coordinator. Christopher teaches units on assessment, digital pedagogies, and leadership across all QUT's education degrees. His research focusses on design-based professional learning and the challenges of integrating digital technologies in teacher practice. He, with Andy Yeh, wrote the Introduction to and Review of this chapter *Design Thinking*.

Andy Yeh is a Senior Lecturer in Mathematics and ICT Education at QUT. His teaching areas include Mathematics Curriculum, Innovative Pedagogy with Digital Technologies and STEM Inquiry Experience. His research is centred on the use, design and evaluation of ICT tools and Makerspace learning environments for STEM education. He, with Christopher N. Blundell, wrote the Introduction to and Review of this chapter *Design Thinking*.

Douglas Hopkos teaches at Bishop Leo Arkfeld Secondary School, Wewak, in the East Sepik Province. He has been a Mathematics and Physics teacher for more than 10 years. He contributed to this chapter *Design Thinking* with a Physics project making use of recycled materials exploring rates of evaporation in the local environment

Lonah Tombele with 19 years teaching experience, is the Head of Technology, Kilakila Secondary School, Port Moresby in the Sabama region. She is currently involved in teacher training and in-service development. Lonah has made contributions to this chapter *Design Thinking in STEM* through a School Conservation Park Project and to Chap. 9 Teacher Leadership outlining her approach to collaborative assessment in STEM.

Manasseh Kui Fena had, at time of writing, 12 years of teaching experience and was teaching STEM Biology at Wawin National School of Excellence, Lae, in Morobe Province. He has since become the Deputy Principal (Administration) at Kopen National School of Excellence in Enga Province. Manasseh has contributed to this chapter *Design Thinking* making use of a recycled laptop battery to generate solar power.

Raphael Ulka is a Mathematics and Computing teacher at Gumine Secondary School, Kudiawa, in the Simbu Province. He has over 10 years of teaching experience and has contributed to this chapter *Design Thinking* with a project designed to solve school water supply problems.

8 Project-Based Learning in STEM

Vinesh Chandra, Dann Mallet, Paula Niandros Polume, Levesa Morumo, Abel P. Mambo, and Elias Wranga

8.1 What is Project-Based Learning in STEM?

Vinesh Chandra and Dann Mallet

Project-based Learning (PBL) is a pedagogical approach that immerses students in real-world and personally meaningful authentic projects that facilitate critical and creative thinking and in the process also develop their collaborative and communication skills. All these qualities are highly valued in the twenty-first century. This approach to teaching promotes active inquiry which can provide a "deeper conceptual understanding that traditional methods of instruction cannot" (Krajcik and Blumenfeld 2006, p. 329).

The PBL approach is often attributed to John Dewey (1859–1952), who believed that education was about preparing learners for the future as a "means to give him [or her] command of himself [herself]; it means so to train him [or her] that he [or she] will have the full and ready use of all his [or her] capacities" (Dewey 1897 as cited in Talebi 2015, p. 4). In the real world, our lives revolve around projects. Each time we engage in a project (e.g. cooking, gardening, jobs around the house), we consciously and subconsciously learn, unlearn, and re-learn (Dewey 1897/2006). Dewey's ideas leaned strongly towards a constructivist approach to delivering instruction. Over the years, the ideas that Dewey proposed have been developed further. For example, Seymour Papert (1928–2016) developed the concept of constructionism, suggesting that students learn best when actively engaged in constructing something (Papert and Harel 1991). All these aspects are fundamental to our STEM Education course design (see Part A). Teaching and learning within the subjects are informed by three main theories—Constructivism (learning by doing), constructionism (learning by building), and social constructivism (learning through collaboration and communication) (Amineh and Asl 2015)—all of which promote Knowles' theory of andragogy (Weegar and Pacis 2012).

Over time, project-based learning has become a widely accepted approach to education with many schools and educators incorporating it into their curriculum and teaching methods. It is often seen

V. Chandra (✉) · D. Mallet · P. N. Polume
Queensland University of Technology (QUT), Brisbane, Australia
e-mail: v.chandra@qut.edu.au

L. Morumo
Wapenamanda Four Square Secondary School, Wapenamanda, Papua New Guinea

A. P. Mambo
Aiyura National School, Kainantu, Papua New Guinea

E. Wranga
Cross Country School of Excellence, Kiunga, Papua New Guinea

V. Chandra et al. (eds.), *Integrated STEM Education in Papua New Guinea*, Sustainable Development Goals Series, https://doi.org/10.1007/978-981-95-0828-0_8

to promote twenty-first-century skills such as critical thinking, collaboration, communication, and creativity, as well as to prepare students for the complex challenges of the modern world. Through projects with contextually appropriate challenges, problems and inquiries can be tackled. In the process, students also learn other skills such as project management. Well-designed projects not only promote learning but they can also prepare students for the real world.

Effective instruction in integrated STEM Education is dependent upon active inquiry (Talebi 2015). Thus, PBL is "well suited as a primary pedagogy for STEM learning" (Odell et al. 2019, p. 2) because it promotes inquiry. For example, by applying the STEM-PBL framework (Forbes et al. 2021), students can tackle real-world challenges through a process that requires them to Ask, Imagine, Plan, Create and Improve to generate feasible solutions. In developing their solutions, students can draw upon their STEM knowledge.

However, linking these connections with the curriculum stipulated by education jurisdictions worldwide is not always easy. In addition, teachers also need to understand STEM concepts before they can visualise connections that exist within projects. Even after teachers have engaged in professional development, how they apply their learnt knowledge and skills can vary. This was evident in a study by Han et al. (2015) who investigated the impact of STEM-PBL professional development sessions that were delivered to 92 teachers over 3 years (10 × 7-h sessions per year). The study specifically targeted the teachers' capabilities of applying their knowledge and skills of STEM-PBL in their classrooms. These researchers concluded that "teachers' understanding cannot guarantee the quality of implementation of STEM-PBL" and how it is enacted in their classrooms (p. 73). They also suggested that teachers' professional learning "should be designed to decrease the gap between the knowing and doing of STEM-PBL" (p. 73). There are other barriers to implementing PBL and connecting STEM disciplines, as well such as standardised testing. For example, Blackley and Howell (2015) believed that, in countries such as Australia, students' performances in standardised tests take centre stage because these outcomes dictate the overall standing of the schools. Further to this, they argued that "the challenge is to reassure teachers that embracing integrated STEM Education and preparation for standardised tests are not mutually exclusive" (p. 112).

Siloed teaching has been a feature of schools for centuries and even in primary schools where there is more scope for integration, "project-based learning is not reflected adequately in school reports and other assessment tools" (Hunter 2020, p.485). Hence, a shift to an integrated approach needs a paradigm and an accompanying need for funding to upskill teachers. Critically, the teaching workforce can only be convinced to make this shift if there is "proof of concept" (Blackley and Howell 2015). There is a lack of quality evidence from developing countries on how integration can be successfully achieved. Given that implementing integrated STEM through PBL is strongly context-based, such evidence is important if teachers are to embrace the shift. Examples of STEM Project-Based Learning (STEM-PBL) are presented through the following four narratives.

TEACHER NARRATIVE 8.1: Paula Niandros Polume led her Grade 9 students at ECOM (Evangelical Church of Manus) Secondary School at Lorengau in Manus Province through a project to provide convenient and reliable access to water in their home communities. The students built (and continue to build) Tippy Taps which make a genuine difference to the health and welfare of their families and neighbours.

TEACHER NARRATIVE 8.2: Levesa Morumo led a collaboration between Grade 9 and 11 students at Kwip Dau Secondary School in the Western Highlands to design and build a prototype of a hydro-electric generator to provide their remote school with a reliable supply of power.

TEACHER NARRATIVE 8.3: Abel P. Mambo has planned a learning experience for his Grade 11 students at Aiyura NSoE which follows the STEM-PBL framework. They will work towards improving the reliability of the electricity supply at the school by installing solar power banks.

TEACHER NARRATIVE 8.4: Elias Wranga planned an integrated STEM project with an entrepreneurial perspective. The students developed a chicken farm along with a cocoa nursery that generated money that goes back into the school. Students learn from what's happening in the farms, through links to the curriculum.

Before engaging with the narratives, it is timely to point out that project-based learning in STEM, while a discrete approach, overlaps with the other approaches presented in Part B. It shares the same intent as the authors, particularly in its focus on active problem-solving and the development of multiple responses to identified problems.

8.2 Teacher Narrative 8.1: Paula Niandros Polume

Paula Niandros Polume

8.2.1 Project-Based Learning in STEM Providing Access to Water

Not all communities in Papua New Guinea have reliable or continuous access to water and this community-based "Tippy Taps" project engaged my Grade 9 students at ECOM (Evangelical Church of Manus) Secondary School to address this very real problem by applying their knowledge in STEM. Tippy Taps are "contactless hand-washing stations" which can be built from local materials. They are easy to maintain and simple to use (World Vision Australia 2023, para. 5). Tippy Taps are typically constructed in developing countries to improve health and hygiene in villages and schools (see Fig. 8.1).

With its obvious concern with public health outcomes, the Tippy Tap project reflects SDG 3 [Good Health and Wellbeing] and SDG 6 [Clean water and sanitation]. With its advocacy of hygiene and proactive engagement with the community, the project has the potential to develop and instil heightened citizenship values in students. Most schools in PNG remain challenged to engage in advocacy programmes in their local communities. The Department of Education's (2020) National Plan indicates the extent of the problem (PNGDOE 2020). My Tippy Taps project hopes to initiate community partnership to extend awareness of the importance of access to water through education.

Completion and installation of Tippy Taps will help develop lifelong skills for self-reliance in students. They will experience a sense of pride in helping their communities by using their skills in Technology and Engineering, their critical and creative thinking and the practical application of Mathematics when designing the Tippy Tap's structure and construction.

8.2.2 Rationale

Project-Based Learning (PBL) is vital for teaching students in the twenty-first century. It is student-centred, requires active participation and motivation not seen in traditional classrooms. It encourages the development of skills such as collaboration, communication, critical and creative thinking, decision-making, time management, public speaking, and organisational and problem-solving skills. Students take ownership and responsibility for their learning as they learn to develop, innovate, and construct products. They can reflect and evaluate as they test their products and continue to make improvements.

Building a Tippy Tap for a specific community demonstrates STEM integration at a classroom level. Students can relate to the need for access to clean running water and, through this project, they will come to appreciate that hand washing reduces and prevents diseases. They will then contribute in small but effective ways to positively impact their communities. Being a meaningful learning activity increases students' engagement (Fortus et al. 2005).

The communities we are targeting are our students' homes. To complete this project, the school moved into a partnership with the

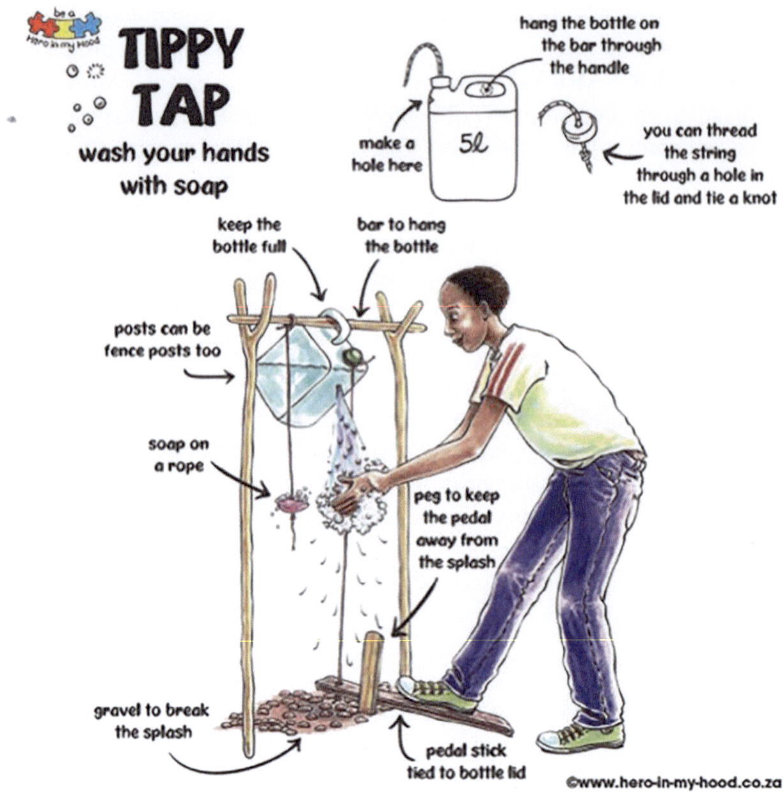

Fig. 8.1 Structure of a Tippy tap constructed in Zambia (*Image Source* https://www.hero-in-my-hood.co.za/free-downloads/tippy-tap/#Community)

local communities to educate through advocacy programmes, to solve daily problems through critical and creative thinking, to collaborate and communicate to build community relations, and to encourage responsible and productive citizenship.

8.2.3 Implementation

The PNG curriculum encourages a student-centred teaching and learning methodology that is tailored for lifelong learning, sustainability, and focuses on integral human development with relevance to authentic learning experiences. It encourages teachers and students to integrate subject knowledge and skills to resolve problems and create solutions for given assessment tasks. It cultivates cultural relevance where students display an understanding and appreciation of the traditional methods of building, counting, measurement systems, trading, communication methods and values that are fundamental to STEM Education. The integrated STEM skills entailed within this small easy-to-build project include:

- engineering skill to plan and construct durable and sustainable Tippy Taps,
- technology skills to use appropriate tools and safety measures when working,
- digital technology skills used for multimedia presentations and submission of typed reports,
- critical and creative thinking skills to design and solve everyday problems at their community,
- mathematical skills to accurately measure the length, width, and height of the Tippy Tap and the length of string and to calculate the volume capacity of the container and the rate of water flow, and
- communication, teamwork, responsible citizen skills to advocate on the health issues faced by their communities,

- STEM thinking skills to systematically engage in research work, resource and data collection and summary, decision-making, giving feedback, testing to solve a problem.

The school subjects integrated in this project are:

- Mathematics Units 9.1–9.4: Create a budget plan, measurements of resources (wood—length, height, width, string—length, container—volume and capacity) get the centre of the lid and soap for hole construction and rate of using tippy tap (PNGDOE 2006).
- Science Units 9.1–9.4: Identify type of resources (wood, pebbles, soap); force applied, applied when using the Tippy Tap. The balance (weight, height proportions of water-filled container) (PNGDOE 2006b).
- Design & Technology Units 9.1–9.4, Engineering: planning and design, structure, construction of project (PNGDOE 2006a).

Table 8.1 summarises the problem statement, the connections to the curriculum and global goals, the pedagogy I chose and a brief reflection on the Tippy Taps project.

The Tippy Taps project was planned to be completed over 10 days with a week given for the completion of presentations. Students worked and were assessed in groups. The project was implemented according to the STEM-PBL model (Forbes et al. 2021) through its five phases of Ask, Imagine, Plan, Create and Improve.

Phase 1 [Ask]: In the first phase, students needed to define and understand the requirements of the project. I provided a task outline and marking criteria. I discussed the project with the students and created groups. I guided them through an exploration of the types of Tippy Taps built across the globe; where are they popular; why are they installed; what resources are used; and the benefits and what are the challenges of constructing them.

Phase 2 [Imagine]: The second phase involved students' investigation and research on the materials they would need to construct their Tippy Taps. They had to explore the stores in their local community to locate the materials needed for the project. I checked the lists of materials and tools they had compiled and commented on the durability and cost of their selections.

Phase 3 [Plan]: In the third phase, the students designed and planned their preferred solutions. Here they had to apply their STEM knowledge, particularly in:

- Mathematics, namely budgets, quantities, and measurement.
- Science, namely the durability of their chosen materials.
- Engineering and Technology, namely structure and placement of the final product.

I provided formative feedback on the students' sketches and the processes of construction they had planned. The members of each group were assigned tasks as part of their summative assessment. I monitored each student's progress on their assigned tasks.

Phase 4 [Create]: The fourth phase is where students constructed Tippy Taps as their final product and solution. Guided by their design plan, they used their selected materials and tools to build their community Taps. They then tested and evaluated their product. I ran tests to assess design in terms of its being user- and eco-friendly, stable, firm and durable.

Phase 5 [Improve]: In the fifth and final phase, students received feedback from peers to evaluate their projects. They were encouraged to use this and the feedback I had provided, to strategies improvements to the current and future construction of Tippy Taps in their local communities. At this stage, all submissions are presented for summative assessment. Students submitted written reports followed by an oral/media presentation or video clip. The presentations were judged by a panel of STEM teachers against a well-defined rubric.

Table 8.1 Narrative 8.1 project summary

Problem	Connections		Pedagogy	Reflection
Access to clean water for hand-washing	*Curriculum Connections*		STEM-PBL (Forbes et al. 2021)	Successful implementation of Tippy Taps in the students' home villages
	Grade 9 Mathematics, Science and Design & Technology			
	SDGs (UN, n.d.)			
	3	Good Health and Well-Being		
	4	Quality Education		
	6	Clean Water and Sanitation		

8.2.4 Outcomes

This project demonstrates the reality of creating a basic, affordable, eco-friendly and safe hand-washing tool as an advocacy programme and solution for minimal spread of contagious hand scabies and COVID-19 in our local community. It can be an advocacy programme with a school-community partnership or by students serving their communities (see Fig. 8.2).

The Tippy Taps were successfully installed in the villages with the help of their respective leaders. The unequal distribution of students into their local level government (LLG) areas was challenging, as some had plenty while others had none. Students living in town were unable to install taps at local primary schools because approval was not sought early enough. These students, however, undertook the project at my residence for assessment.

8.2.5 Reflection

The construction of Tippy Taps is an example of project-based learning in STEM that allowed students at my school to address a community-based issue, that is, providing a convenient access to water for hand-washing. The taps provide visual reminders to wash your hands for good health and hygiene and to offer protection at critical times such as outbreaks of scabies and the COVID-19 pandemic. This project makes a contribution to the achievement of the SDGs: SDG 3 [Good Health and Well-Being], SDG 4 [Quality Education] and SDG 6 [Clean Water and Sanitation] (Fig. 8.3).

The benefits of the project include transferring of lifelong skills through authentic learning and applying critical and creative thinking to solving everyday problems with their communities. Students also develop a sense of responsibility as citizens or as members of their communities and learn to appreciate values, virtues, and attributes of good citizenship. Through interdisciplinary STEM, the project aimed to emphasise the importance of twenty-first-century skill development such as adaptation, social skills, communication skills, problem-solving and self-development (Bybee 2010).

Most importantly, the students got to see the practical applications of STEM subjects, making the student more receptive to these subjects taught in class. STEM gives an opportunity for students to understand real-world problems based on those interdisciplinary subjects (Dugger 2010). There was an intrinsic reward in doing something positive for their community.

Now that we are experiencing a dry spell in our school, I have decided to reintroduce the project with the Grade 11 students to help with hand washing after the use of our bush material toilets (pit toilets) where we do not have water taps available. The project has an important role to play in the health and hygiene of everyone in our community.

Fig. 8.2 Completed Tippy Tap project in Timonai Village

Fig. 8.3 SDGs met through a community project

8.2.6 Postscript

By 2025, the Tippy Tap project had become a common community project. The Grades 11 and 12 students from ECOM Secondary School go home to their respective villages in the term breaks where they build the taps for their primary or elementary schools with the help of the village leaders. The purposes have remained the same and this authentic task has continued to meet local and global goals for my students and their communities.

8.3 Teacher Narrative 8.2: Levesa Morumo

Levesa Morumo

8.3.1 Project-Based Learning in STEM Improving School Electricity Supply

Electricity supply through PNG Power is inconsistent and outages are frequent. In many remote schools, like Kwip Dau Secondary School, there are frequent disruptions to student learning when power and lighting are interrupted. The school computer lab, photocopiers and night study areas are all affected. As the school is located away from the main township, electrical faults are not rectified quickly, and blackouts can persist for some time. Not having access to basic services hampers attempts to improve the quality of education throughout the country.

Working with two classes—Year 9 and Year 11—my colleague, Jimmy Tai and I decided to investigate a local solution to the problem. This investigation would pose an authentic problem for students to investigate and would allow us to meet the respective requirements of the Grades 9 and 11 Science and Mathematics syllabuses. It gave the opportunity for student collaboration across and within year levels. We believed that adopting project-based learning was the best approach to take because, while we knew the outcome we were after, there were multiple ways it could be achieved.

8.3.2 Rationale

The PNG government is focused on improving the quality of education at all levels but is hampered by problems in basic service delivery throughout the country (Asian Development Bank 2008). For example, the lack of a consistent and reliable electricity supply, particularly for remote schools, poses challenges for setting up computers and accessing digital services which have become critical in the past few decades (Department of Information and Communications Technology 2023). It also causes disruptions for students' learning. This project looks at one school's response by addressing this issue.

According to Capraro et al. (2013), project-based learning is "an ill-defined task with a well-defined outcome situated within a contextually rich task requiring students to solve several problems, which when considered in their entry, showcase student mastery of several concepts of various STEM subjects" (p. 4). For example, students can have hands-on experience integrating all STEM disciplines. Through this process, they can develop lifelong skills. Tackling "ill-defined task(s)" (Capraro et al. 2013) promotes creativity and innovation.

Having access to a small hydro-electric generator to provide reliable power for our remote school will help reduce the gap between us and metropolitan schools. For our students, building a generator will allow them to address real-world problems and enhance their STEM competencies and skills. This hydro-electric project also facilitated collaboration between teachers who usually teach STEM through a siloed approach. As a multidisciplinary task, teachers can work collaboratively in assessing the different components of the task across the subject as there is a link between the subjects. Table 8.2 summarises the problem statement, the connections to the curriculum and global goals, the pedagogy I chose, and a brief reflection of the project.

8.3.3 Implementation

The first step in the STEM-PBL framework (Forbes et al. 2021) is to identify the challenge or issue to be investigated. For us, as noted, it was the school's unreliable electricity supply which caused major disruptions to daily life.

The second step is to establish the connection between the project and the curriculum. There were a number of clear connections within the project to the Grade 9 and Grade 11 STEM subjects our respective classes were studying (Table 8.3). These can be summarised as follows:

The third step is concerned with assessment. Before beginning the project, we involved the students in writing the standards for the project's assessment rubrics. Rubrics help teachers to evaluate students' learning efficiency (Andrade 2000) and let students know what is expected of them. It also provides guidelines for students through the peer and self-assessment (Andrade 2000). The criteria we used were Knowledge, Skills, Testing, Prototype Designing and Building.

The fourth step, the teaching and learning process in the STEM-PBL framework, has five phases. These are: Ask, Imagine, Plan, Create and Improve.

Table 8.2 Narrative 8.2 project summary

Problem	Connections		Pedagogy	Reflection
Continuous power outages	*Curriculum Connections*		STEM-PBL (Forbes et al. 2021)	Authentic real-world experience. Successfully generated 5–7 V through a small hydro-electric generator
	Grade 9 Science and Mathematics Grade 11 Physics, General Mathematics and Design & Technology			
	SDGs (UN, n.d.)			
	4	Quality Education		
	7	Affordable and Clean Energy		
	8	Decent Work and Economic Growth		
	9	Industry, Innovation and Infrastructure		
	10	Reduced Inequalities		

Table 8.3 Curricular connections

	Science	**Design and Technology**	**Mathematics**
Grade 9	Science (PNGDOE 2006d) Unit 9.6: Generating electricity	Investigating Designing Creating Evaluating	Mathematics (PNGDOE 2006c) Unit 9.1: Mathematics in our community (Measurement & ratio)
Grade 11	Physics (PNGDOE 2008b) Unit 11.1: Work power and energy Unit 11.5: Electricity		General Mathematics (PNGDOE 2006b) Unit 11.1: (Numbers and Applications). Topics on mass, volume, and pressure

Phase 1 [Ask]

The Year 9 cohort is divided into four classes, each of 30 students. For their part of the project, they were asked to work in groups of 10 to plan a renewable source of electricity production for our school.

The Year 11 cohort has 50 students. In their science class, they were shown a video on how to generate electricity. Then they were organised into five groups of 10 and asked to identify renewable sources of electricity and how it can be produced. They were asked to review the Grade 9 designs and improve on them by creating a prototype. Both Grades 9 and 11 students are aware of the problems with the electricity supply in the school. They also know that the project we are initiating is significant which Williams (2019) defined as being "useful to the student, having personal meaning, being purposeful and intentional (as distinct from random and meaningless), productive and integrative" (p. 020,001–4). This allows students to think critically considering the context and the availability and suitability of resources. In short, the Grade 9 students came up with the plan and design and presented their findings for assessment. These were given to Grade 11 for adjustment and creation of a prototype. This narrative describes how this multi-grade activity worked with a small group.

Phase 2 [Imagine]

Students from both cohorts will work in groups to carry out an investigation of electricity, its types and means of generation. They will consider the effect of topography and geography on the generation and transmission of electricity considering our school at Mt Hagen. Students will also be asked to identify the materials that could be used for creating a prototype.

The Grade 9 students will carry out an investigation on electricity based on stimulus questions such as:

- What causes the light to glow?
- What are sources of electricity?
- How can electric current be produced?
- What are the renewable sources of energy used to produce electricity?

They were then asked to choose one renewable source of energy and design a plan for how it could be harnessed to produce electricity.

Year 11 students were presented with the Grade 9's plans for energy generation. They were allowed to alter the plans or, if they were satisfied, to work with what they were given. They did independent research into types of renewable energy and how to generate electricity using renewable sources. They also identified the materials to be used for creating the prototype. Based on the geographical location of the school, they will critically analyse and identify suitable methods of electricity generation to be enacted with the prototype.

Phase 3 [Plan]

Based on their research findings, each cohort was asked to explain how they intend to carry out their project through discussion among themselves. The Grade 9 students presented and explained their findings and plans in their groups. This allows for peer and self-assessment to critically analyse the plan. In doing so, they noted what needed to be included in their plans and outlined the materials that could be used to create the prototype.

The Grade 11 students were asked to explain how they intended to carry out the project, based on their findings and the plans supplied by the Grade 9 groups. Through discussion among themselves, they identified the materials to be used in creating the prototype.

Phase 4 [Create]

Students will work on planning and building their prototypes. The Year 9 groups will conduct research and produce plans for a renewable power generator to solve the school's energy supply problems. Figure 8.4 is a Year 9 schematic diagram for a hydro-electric generator.

The group responsible for the sketch in Fig. 8.5 suggested that a prototype could be built using a mixture of new and recycled materials: butter lid, motor, fan belt, metal rod, plywood, biros, plastic lids from soft drink bottles, pipe, bucket, container and liquid nail glue. This group presented their plan to the class and were assessed based on an agreed rubric.

A Grade 11 group took the Grade 9 sketch and made adjustments for their prototype. Figure 8.5 shows one of their revised sketches.

The Year 11 students also identified further materials that could be used to create their prototype. After finalising their plans and collecting the desired materials, the prototype was created (see *Outcomes*).

Phase 5 [Improve]

The students will reflect on the functionality and operations of the prototype they created. They will use trial-and-error to improve their design. The fifth step of the STEM-PBL framework is Reflection. Students review and reflect on their learning as to whether they have understood the concept and the connection to the different fields in STEM. For the Grade 11 students, design and construction can also be evaluated considering the factors affecting the power output (height of the output water, the motor used, angle and rotation of the turbine and the vacuum for water circulation). Based on the rotation of the motor and power

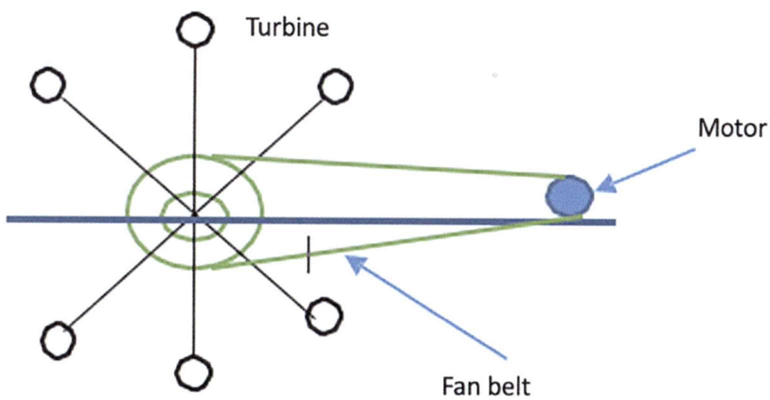

Fig. 8.4 Grade 9 sketch for a hydro-electric generator

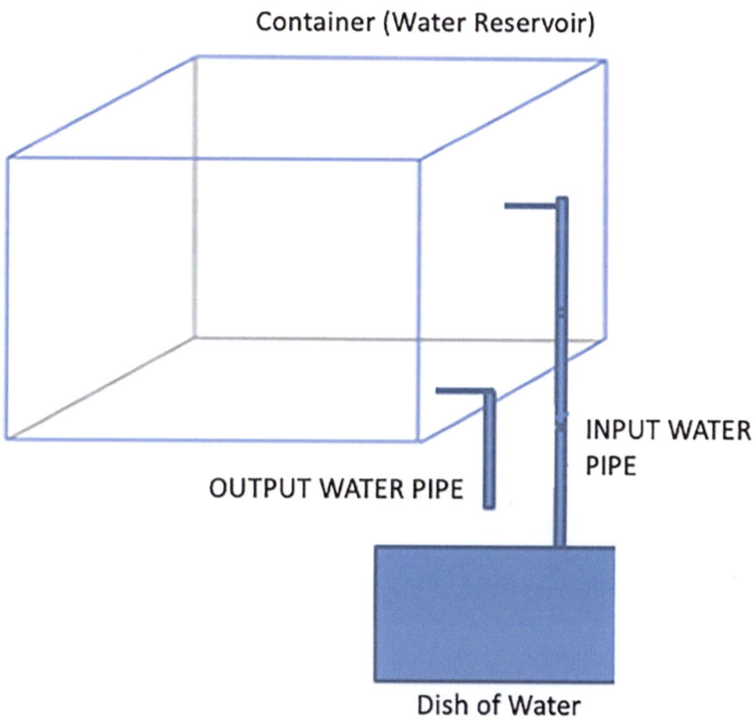

Fig. 8.5 Grade 11 sketch for components of planned hydro-electric generator

output evaluation can be made for improvement. Evaluation of the Grade 9 students can be based on the sketches they produced and their capacity to explain the underlying processes.

8.3.4 Outcomes

One group of Year 11 students was able to construct a working prototype of a hydro-electric generator for the school. They worked, as planned, from the sketches created by Year 9 students. Figures 8.6 and 8.7 show the completed prototype.

Figure 8.8 shows a test we conducted to generate sufficient power to charge a mobile phone. Water from the container was funnelled through the pipe which created sufficient pressure to turn the turbine attached to the DC (direct current) motor to produce 5–7 V DC to charge the phone. The motor used for the prototype produced 12 V of direct current. The rotation of the turbine produces electricity. What we did was simply reverse the motor to become a generator by using

Fig. 8.6 Prototype showing position of motor and turbine

the rotation of the turbine to produce electricity. Now we use rotation to produce electricity. The usefulness of this model was to create the means to generate electricity. This model project could be explored further to develop micro-hydropower for household items (basic motors). In this way, standard 240 V AC (alternating current) could possibly be harnessed in remote schools where electricity is an obstacle to teaching and learning.

8.3.5 Reflection

Challenges were met through the project. Implementation was difficult, particularly in meeting time-bound curriculum demands. The project needed double lessons as students need extra time in class to execute their plans with supervision and gain hands-on experience. More time was needed for the in-depth research the students needed to do.

Despite this, there were many benefits from the prototype project. This includes teaching and learning, authentic real-world experience and cost effectiveness. Enthusiasm and engagement were created, unlocking students' creativity. Learning outcomes, described in the respective syllabuses, were met through the project. As a result, students drew knowledge from the STEM disciplines as an integrated interdisciplinary approach. They used their mathematics knowledge through measurement and ratio, and calculations of mass, volume and pressure. They used their science knowledge regarding electricity generation, work and power, fluids and electromagnetism. Their experience of technology proved critical when it came to the use of tools and equipment while engineering skills were put to use in the design and construction stages.

Students "learnt by doing" in this project facilitated by their teachers. They learnt from each other both within and between grades. The formative peer and self-assessment we adopted provided valuable opportunities for learning. Involving students in the development of the rubrics we used for their summative assessment was key

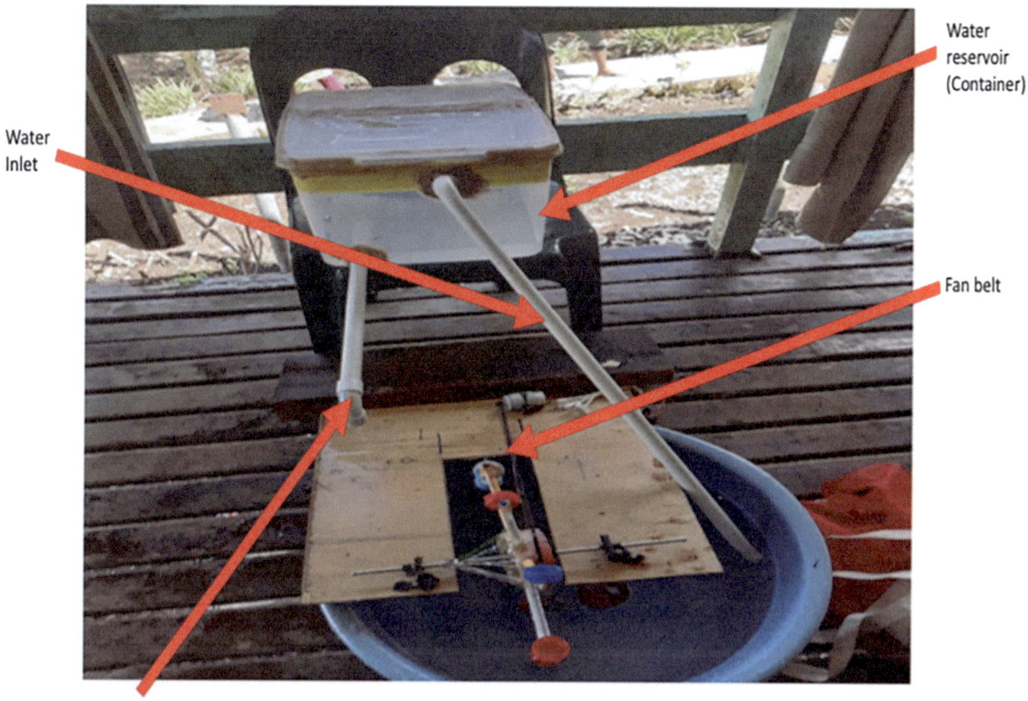

Fig. 8.7 Prototype showing PVC inlet and outlet pipes, water reservoir, and fan belt

in students understanding what was expected of them.

In a small way, this project contributes to the achieving the goals of the PNG Medium Term Development Plan III (Independent State of Papua New Guinea 2018) and Vision 2050 (Independent State of Papua New Guinea 2010) through implementing PBL in STEM Education. It also works towards the achievement of SDGs (Fig. 8.9). It directly addresses the SDG 4 [Quality Education] in adopting a project-based learning approach and SDG 7 [Affordable and clean energy] by introducing students to the principles of electricity generation. It will also indirectly address the SDG 8 [Decent work and economic growth] and SDG 9 [Industry, innovation and infrastructure] because of the productive and transferable skills the students will develop. The certainty of a reliable power supply will also address SDG 10 [Reduced Inequalities].

The project-based learning strategy we adopted gave our students a structure to follow to work towards an ill-defined solution to a well-defined authentic problem. We found that we can produce affordable renewable power and energy to supplement education and improve living standards in our community while progressing our students' STEM knowledge.

8.3.6 Postscript

I now teach at Wapenamanda Four Square Secondary School in Enga Province. While at Kwip Dau Secondary School, my students completed the project and then continued to make improvements with the prototypes they had built. They also ventured into looking into electromagnetic induction, an important aspect in hydroelectricity generation.

Fig. 8.8 Prototype testing

8.4 Teacher Narrative 8.3: Abel P. Mambo

Abel P. Mambo

8.4.1 Project-Based Learning Through a Solar Power Bank from Recycled Laptop Battery

As for other schools in PNG, an unreliable electricity supply is a problem we experience at Aiyura NSoE (see also Narrative 7.3 Manasseh Kui Fena). Due to power blackouts and fluctuations, computers and printers are affected, and studies are disrupted. The school spends money fixing damaged items and paying power bills. My solution was to have Grade 11 STEM students, in collaboration with STEM teachers, build a solar power bank.

The project would use solar panels and lead accumulators to convert solar energy from the sun into electrical energy. Table 8.4 summarises the problem, the curriculum connections, pedagogy and a reflection on this project.

8.4.2 Rationale

Moore et al. (2014) defined integrated STEM Education as "an effort to combine some or all of the four disciplines of Science, Technology, Engineering, and Mathematics into one class, unit, or lesson that is based on connections between the subjects and real-world problems" (p. 38). Brophy et al. (2008) argued that STEM-PBL allows students to connect relevant skills in real-world applications by providing valuable learning contexts. Similarly, Tseng et al. (2011), using the acronym, PjBL, found that the close relationship

Fig. 8.9 SDGs met through an electricity production project

Table 8.4 Narrative 8.3 project summary

Problem	Connections		Pedagogy	Reflection
Unreliable electricity supply that affects technology and disrupts its use during school	*Curriculum Connections*		STEM-PBL (Forbes et al. 2021)	Practical issues for the project's implementation have been identified and resolved
	Grade 11 Applied Science, Design & Technology, Engineering and Maths			
	SDGs (UN, n.d.)			
	3	Good Health and Well-Being		
	4	Quality Education		
	7	Affordable and Clean Energy		
	10	Reduced Inequalities		
	13	Climate Action		
	17	Partnerships for the Goals		

between the STEM disciplines and the integration of these subjects can help students develop relevant knowledge, concepts, and skills.

The school operates with an overarching aim to have good results at the end of the academic year. To achieve this aim, students must fully concentrate on their studies. However, studies are interrupted due to continuous power blackouts and fluctuations. The fluctuations have also destroyed some high-cost machines like printers and computers.

The school has a standby generator that goes on during the blackouts but due to the current economic drop (inflation) of the country the prices for the goods and services are "skyrocketing". Prices for fuel are very high and, sometimes, fuel cannot be delivered to the school due to natural disasters affecting the road.

In response, students at Aiyura NSoE will build a solar power bank. It relates to the PNG curriculum and SDGs. It will supply enough power to the school depending on the number of a lead accumulator and solar used. The use of power well will be sustainable and cheap.

8.4.3 Implementation

This project will follow the STEM-PBL model devised by Forbes et al. (2021). Table 8.5 summarises the three steps needed to prepare for this project (problem identification, curriculum connections, assessment schedule) before outlining the five phases in the teaching and learning process (ask, imagine, plan, create, improve).

Table 8.5 Narrative 8.3 PBL project preparation

Preparation (Steps 1–3)	*Problem Identification*	A reliable electricity supply is needed at the school. Solar power banks may provide a viable renewable solution to our power supply issues. (See *Step 1* below)
	Curriculum Connections	Students will use all STEM disciplines in this project. Knowledge and skills will be drawn from Grade 11 syllabuses. (See *Step 2* below and Table 8.2)
	Assessment Schedule	The timing and nature of formative and summative assessment is planned
Teaching and Learning (Step 4)	Phase 1–Phase 5 (Forbes et al. 2021)	
Reflection (Step 5)	Looking back	Reflect on what worked, what were the challenges and how could the activity be improved

Step 1 Problem Identification

Electricity is the problem in this case. Continuous blackouts and power fluctuations affects students' academic performance and causes damage to school equipment. Solar power banks will be investigated as a solution to the problem.

Step 2 Curriculum Connection

The PNG syllabus, guides the teaching, construction and implementation of the solar power banks. Table 8.6 outlines the subjects, topics and skills used.

Step 3 Assessment Schedule

Students will be assessed using both formative and summative assessment through the five weeks of the project. Formative assessment will monitor student's weekly progress. Photo journals are to be kept in all the process. Summative assessment will assess students' overall individual performance through a final report and an oral presentation. The presentation will be judged by an independent body using a rubric.

Step 4 Teaching and Learning Process

Teaching and learning processes are described as five iterative phases.

- *Step 4* **Phase 1 [Ask].** Students are introduced to the project goals and requirements. Task outlines and rubrics are provided and discussed. Small groups are formed based on the number of students. Teachers check for understanding of the problem.
- *Step 4* **Phase 2 [Imagine]:** This phase has students investigating and researching the project. They explore and enquire on the internet or ask the people around them to solve the problem. Consultation can be done with the school administration to provide funds to purchase the materials that will be used in the project. The teacher helps the students to ask for funds so that appropriate materials are available.
- *Step 4* **Phase 3 [Plan]:** In this phase, students design and plan their preferred solution. STEM skills and concepts are applied. Science concepts are applied to identify the materials to be used; Mathematics is applied to measure and calculate the cost; Technology is applied to identify the tools needed and analyse data; and Engineering is applied to identify the safety measures and precautions.
- *Step 4* **Phase 4 [Create]:** Students now produce their preferred solution using the plan they developed in Phase 3.
- *Step 4* **Phase 5 [Test]:** In the final phase, the solution is tested. This determines whether to improve or approve the product. If the test is unsuccessful, students return to Phase 1 [Ask]. This time, the students' questions

Table 8.6 Narrative 8.3 PNG curriculum connections

Subject	Unit/Topics	Connection to Solar Power Bank Project
Science–Physics (PNGDOE, 2008b)	Measurement • Quantities and units • Measuring instruments • Error analysis	Measurement of length and width; type of measuring instruments to be used
	11.4 Work, Power & Energy • Energy • Power • Simple Machines	Types of energy, particularly solar energy
	11.5 Electricity Principles • Current electricity • Wheatstone bridge and potentiometer • Alternating current (AC) circuits	Electricity concepts, particularly connecting the batteries and serial or parallel electrical applications
	11.6 Electronics • Solid state electronics • Digital electronics	Electronics concepts particularly is diodes and circuit board. Ohm's Law used to calculate resistance, voltage and current
Mathematics—General Mathematics (PNGDOE, 2006c) (PNGDOE, 2006b	11.4 Geometry • Basic numeracy • Units of Measurement • Ratio and proportion • Basic algebra	Mathematics is applied to calculate: • the area of the circuit board • the shape of the circuit board comes • the angle in which the solar bank will be placed to attract maximum sunlight
Technology—Design & Technology (PNGDOE, 2008a)	• Design • Data Analysis • Construction Technology • Building Technology	Technology tools are used to build, construct and test the solar power banks
Engineering—Design & Technology (PNGDOE, 2006a)	• Product design • Prototype • Testing • Biomimicry Design Process • Occupational Health and Safety	Engineering concepts will guide the development of the solar power banks

seek to identify the cause of the failure or unexpected performance before the cycle resumes. However, if the test is good, it is presented to the teacher for assessment. It may also be presented to others, such as the school administration, for future implementation.

Step 5. Reflection

The final step in the STEM-PBL Model is reflection. I will document the project carefully so that I can improve on later iterations or to advise teachers in other schools who are interested in solar technology or renewable energy projects.

8.4.4 Outcomes

On a practical level the outcomes I have identified in this project planning are: resources, installation, budget and time.

- *Resources*: I envisage that resourcing will be a challenge for this project and will be a major factor in determining its success. For instance, circuit boards and solar panels cannot be easily found. Other materials like connecting wires and lead accumulators (batteries) can be recycled from old or damaged radios, electrical appliances and laptops. Inactive car batteries can also be used. Solar power bank kits are available commercially but are expensive and have to be sourced from outside PNG.

- *Installation:* Solar panels generate less electricity during cloudy or rainy days. Thus, the normal flow of the functioning of the school could be affected by the weather. Combining solar energy with other renewable energy sources (like wind and hydro-electricity) will need to be investigated to ensure a constant supply of electricity.
- *Budget*: An approach has been made to the School Principal for funding for this project. I have made the argument that installing solar power banks will save the school money by preventing damage to electrical infrastructure from blackouts and power fluctuations. Once a budget has been established, it will be critical to not exceed spending.
- *Time*: The project must be completed within the time frame given. Last-minute work will affect the final result. Work allocated per week must be done.

I have also had a chance to consider three potential benefits for the implementation of solar power banks in the school. The first benefit is that continuous light and power will help students to focus on their studies without disruption. Studies will not be jeopardised, assessable tasks are done on time, and they will achieve good results. Students will be encouraged to consider the potential benefits to life in PNG, particularly in remote villages.

The second benefit is an unexpected opportunity created by the necessity to use recycled materials (because of funding constraints). That is, that students will be forced to become creative thinkers and problem-solvers. This resilience will be a useful and transferable skill for life after school.

The third benefit is environmental. Solar energy is renewable and sustainable resource of energy which means that it can be used without harming the environment. Using solar power banks will contribute towards a decrease in global warming due to the burning of fossil fuels. During electricity blackouts, the school's generator goes on automatically, if fuel is available. It produces a lot of carbon dioxide (CO_2) which is dangerous to the world.

The project has the potential to address broader issues, that is, to work towards achieving the SDGs (see Fig. 8.10). These are SDG 3 [Good Health and Well-Being], SDG 4 [Quality Education], SDG 7 [Affordable and Clean Energy], SDG 10 [Reduce Inequalities], SDG 13 [Climate Action] and SDG 17 [Partnerships for the Goals].

8.4.5 Reflection

This project aimed to engage STEM students in exploring locally available renewable materials. It hopefully helped them to see the connections between the STEM disciplines they are studying. I also hoped that they will develop twenty-first-century skills including critical and creative thinking, and skills in engineering, technology and digital fluency. Finally, I hoped the students will appreciate the broader goals of this project as expressed in terms of the SDGs designed to ensure a peaceful and healthy life for all.

8.4.6 Postscript

This project might not have been a success because we were unable to complete it in the time available. My students collected some power banks (lead accumulator), which will be used by my current cohort of Grade 11 students. The project, apart from being useful and finding a solution to a very real problem, provides an excellent way to integrate STEM disciplines and enhance student knowledge and skills.

Fig. 8.10 SDGs met through a solar power bank project

8.5 Teacher Narrative 8.4: Elias Wranga

Elias Wranga

8.5.1 Project-Based Learning in School-Based Entrepreneurship Culture for Learning and Earning

STEM Education has been adopted in many countries including Papua New Guinea (PNG) to engage students in real-world rigorous learning. The culmination of STEM Education is innovation and enterprise and therefore the schools need to balance the curriculum's academic work with the rigorous real-world experiential learning to cultivate the skills and attitude of being innovative and enterprising (Mbanefo and Ebokab 2017). The goal of every innovator is to become an entrepreneur and for every entrepreneur to become an innovator (Elbaz et al. 2013). STEM Education is an enabler of this journey for aspiring students to be future innovators and entrepreneurs (Camesano et al. 2016). Learning is not just knowledge acquisition but learning how knowledge can be useful beyond theory and reaching the knowledge economy and commercialisation stage. This will inspire students' entrepreneurial thinking that will drive creative and innovative solutions while becoming strong communicators (Doga-Mîrzac 2017). Furthermore, it is likely that the engineering and science students of today are preparing for jobs that do not currently exist. Having a firm grounding in entrepreneurship with an emphasis on innovation and invention may be the best way for these students to prepare for the twenty-first-century workforce (Camesano et al. 2016).

The development of a STEMpreneurship Culture is a strategy to consolidate integrated STEM learning using project-based learning (PBL) in schools whereby specific measurable projects are designed within the school curriculum context to reach the national learning standards. It also embraces the dual nature of learning and earning simultaneously where teachers can measure their (students) learning effort through multifaceted assessment and income or expenses. Several projects can be designed and implemented in harmony without disrupting teaching periods, timetables or other extracurricular school activities through system thinking approach (Cabrera et al. 2018).

Rapid shifts in industry (Fourth Industrial Revolution) and the workplace means that schools globally are required to align learning content with the development of relevant and versatile knowledge set, skillset, mindset and toolsets so that graduates remain competitive on the job market or being entrepreneurial and employable (Lewis 2005). Such paradigm shifts in teaching and learning can be successfully underpinned by PBL that interests and motivates students to illustrate core knowledge relating to the real world (Curtis et al. 2023; Pringle et al. 2020).

This project works towards the achievement of the SDGs (Fig. 8.11). These are SDG 1 [No Poverty], SDG 2 [Zero Hunger], SDG 3 [Good Health and Well-Being], SDG 4 [Quality Education], SDG 8 [Decent Work and Economic Growth] and SDG 11 [Sustainable Cities and Communities].

8.5.2 Rationale

In PNG, government funding of schools is insufficient. In many instances, it is not sufficient to support students' learning needs. Underfunding impedes effective implementation of STEM learning in all schools annually. Therefore, initiating a *STEMpreneurship* culture at the Cross Country School of Excellence is a way forward for steady income generation, skills development, innovation and enterprise. This was recently done with the launch of a cocoa nursery and chicken farm project running in parallel. Figure 8.12 shows students at work in the cocoa nursery and budding cocoa plants.

Incorporating entrepreneurship activities in STEM Education will hopefully prove to be a sustainable approach in learning and earning simultaneously. The overarching goal of developing a STEMpreneurship culture in the school is to enable students to explore the intersection of STEM and entrepreneurship in hands-on-project-based learning where entrepreneurship skills are imparted at a very young age (Kärkkäinen and Lancrin 2013). The diagram (see Fig. 8.13) demonstrates a synergy of linear and cyclic systems thinking as mechanical gears. School children can be engaged in the learning process whilst the same learning is an avenue for generating income from what they are learning about.

The STEMpreneurship culture of the Cross Country School of Excellence is underpinned by the second goal of PNG Vision 2050 for "Wealth Creation" (Independent State of Papua New Guinea (2010)). This part of the PNG Vision also aligns with the Sustainable Development Goals of Quality Education (SDG 4) and Decent Work and Economic Growth (SDG 8) (UN, n.d). Adopting new and better ways of learning will help today's children in their future. The STEMpreneur project is also ideal for teachers to engage students through project-based learning giving both teachers and students the opportunity to engage in a hands-on activity that iterates both teaching and learning, respectively. An additional benefit is that some projects have the potential to earn income which can contribute towards the sustainability of learning at the Cross Country School of Excellence. Using both Systems Thinking and PBL has proven to be very practical for classroom brainstorming and outdoor experiential learning. Table 8.7 summarises the project in terms of its problem identification, its connection to curricular and global goals, the chosen pedagogy and a brief reflection on the outcomes achieved.

8.5.3 Implementation

A STEMpreneur cocoa nursery and chicken farm integrated farming project has been undertaken by students from kindergarten to primary level with ages ranging from 5–13 years old. The project is based on the school grounds where progress is constantly monitored. The vision is for STEMpreneurship Culture to become an avenue for experiential learning that enables students to acquire knowledge, skills and income to build their innovative and enterprising attitude for life. This will be achieved by designing more STEMpreneurship-based projects aligned to the school curriculum standards and benchmarks while also collaborating with other teachers, parents, community members and industry to provide necessary resources, skills and support.

The framework of a STEMpreneur project can be supported or managed by the community members or handymen engaged by the school. The Systems Thinking applies S.M.A.R.T goals for the STEMpreneur projects to operate within the parameters as they are connected in linear

Fig. 8.11 SDGs met through an entrepreneurial integrated farming project

or cyclic systems for harmonious operation. The Inputs, Processes and Outputs are appliable to most STEMpreneur projects. However, changes are made to suit specific STEMpreneur projects. Students are involved in all three stages of the STEMpreneur Project (see Fig. 8.14).

The purpose of the cocoa nursery and chicken farm project is to provide hands-on learning projects that will achieve the curriculums' aims and generate income for continuous STEM learning. The project adopted the following specific goals to be completed by the end of term.

- The purchase of 10 laptops for ICT lessons.
- Set up the benches for the computer lab.
- Purchase one RACHEL WiFi technology.

The STEMpreneur project included a broad range of collaborators including the teachers, students, school handymen, community people, technical experts and industrial partners. The project used SMART (Specific, Measurable, Achievable, Reliable and Time-bound) goals to operate a cocoa nursery and chicken farm in parallel and generated 10,000 K. The project was managed by the teachers, students and the school handyman. The project is not so scientific or too technical in nature that it required specialised knowledge and skills or expertise. Besides, it is within the funding budget and timeframe at scale for a school term. Financially, the expected return on investment was greater than the expense, so it fell within the operational budget. Both projects were implemented by the end of the term and began generating income.

8.5.4 Connections to the PNG National SBC Curriculum: Primary School

The project-based learning is driven by a multidisciplinary approach of science, social science and math in the PNG Standards-Based Curriculum to integrate the identified skills and knowledge. Based on this transdisciplinary connection, the STEMpreneur PBL has been derived to complement the content knowledge, impart the skills, and develop an attitude. The content teaching approach is 'Sage-on-the-Stage' while the PBL session will be delivered through 'Guide-on-the-Side' or 'Scaffolding Method'.

The targeted Science Standards are across Grades 3, 4, and 5. Strand 1: Life, Unit 1: Plants. Learning Standards 3.1.1, 4.1.1, 4.3.1, and 5.1.1 aim to provide students with practical experiences in understanding the importance of sustainable development practices in their own environment and experiences which build on their interest in and stimulate their curiosity about their

Fig. 8.12 Students engaged in cocoa production

environment (PNGDOE 2017b). In Plant Science, students will:

Learning Standard 4.3.1: Investigate and explain the importance of soil for living things.

Practical implementation: Students collected humus soil, studied its properties, and compared with other soil types. They filled up the poly bags in preparation for nursery.

Learning Standard 3.1.1, 5.1.1: Parts of plants and needs of plants. Observe the stages of plant growth over a period and provide their necessary needs for healthy growth.

Practical implementation: Students prepared cocoa seeds and germinated them

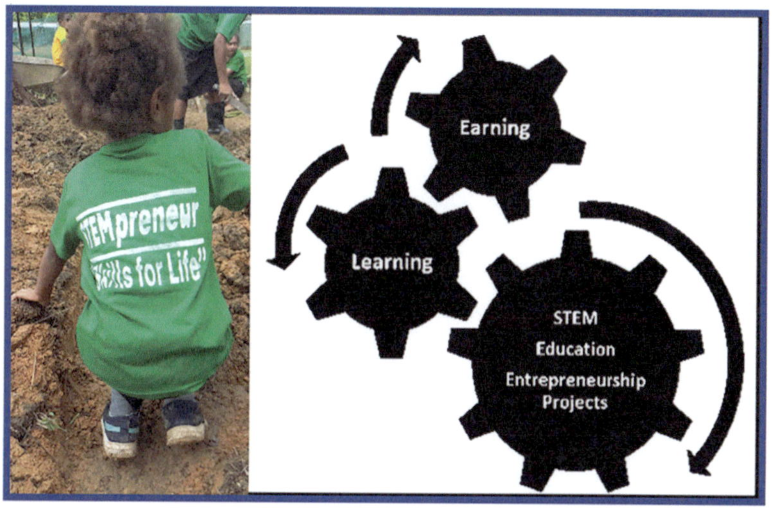

Fig. 8.13 The STEMpreneur culture: Self-sustainable harmonious operation

Table 8.7 Narrative 8.4 project summary

Problem	Connections		Pedagogy	Reflection
School income from government is both late and insufficient	*Curriculum Connections*		5E inquiry Framework (Bybee 2010) Entrecomp Framework (Bacigalupo et al. 2016) Visible Learning (Hattie and Yates 2014)	Chicken farm and cocoa nursery produced K16000 that was used to purchase computers and other technologies to support ICT teaching
	Science (Plant and Animal), Maths			
	SDGs (UN, n.d.)			
	1	No Poverty		
	2	Zero Hunger		
	3	Good Health and Well-Being		
	4	Quality Education		
	8	Decent Work and Economic Growth		
	11	Sustainable Cities and Communities		

externally. After three days, the germinated seeds were transferred to the poly bags. They placed the seeds correctly in the soil so that shoots grow up for sunlight (phototropism) and tap root grows down for water and nutrients (gravitropism and hydrotropism). They then place the poly bags in a shady place and water them.

Learning Standard 4.1.1: Describe the life cycle of plants.

Practical implementation: Students observe the seedling, observe the matured cocoa trees with flowers, young cocoa pods, matured pods, and colours of the ripe pods; and,

Learning Standard 4.1.1: Sketch the growth of plants at each stage of its life cycle.

Practical Implementation: Create charts for the cocoa lifecycle and do an oral presentation.

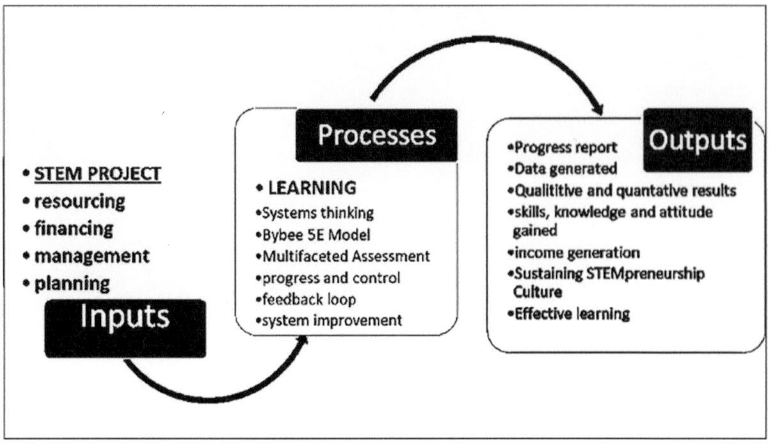

Fig. 8.14 Thinking in STEM

Learning Standard 5.1.1: Identify and state the basic needs for plants to grow.

Practical implementation: Students studied the seedlings under different conditions: (i) Conducive condition with all the needs of plants, (ii) no water, (iii) total shade, and (iv) a totally dark room. They record and present what they have observed.

In Animal Science, the standards are across Grades 3, 4 and 5. Unit 1 Animals learning strands: 3.1.2, 4.1.2, 5.1.2 aim to investigate and explain body parts of animals and state their functions, identify and explain their different stages in the life cycle of the animals from reproduction to birth and adult.

Learning Standard 3.1.2: Investigate and explain animal body parts and their functions.

Practical Implementation: Students studied the external parts of the chicken, and their functions then move on to dissect the chicken to investigate the internal organs and their functions. They then use the chicken meat for their nutrition cooking in Health and Nutrition lesson.

Learning Standard 4.1.2: Identify and explain the different stages in animal life cycles.

Practical Implementation: Study the life cycle of the chicken.

Learning Standard 5.1.2: Understand animal reproduction to ensure continuity.

Practical implementation: Investigate the mating process and reproductive organs of the chicken and the development of the eggs.

The targeted Social Science Standards are across Grades 3, 4 and 5 (PNGDOE 2017c). Unit 2: Trade. Learning Strand 2: Production. Learning standards 3.2.2, 4.2.2, 5.2.2 aim to provide students with: The knowledge of cash crop. Investigate and describe how a cash crop is grown, harvested, and processed. Explain how goods are produced from food and cash crops and are traded for income.

Learning Standard 3.2.2: Describe the process for making products that can be sold for cash.

Practical Implementation: Students ferment the cocoa beans and produce homemade milo and chocolate. The students showcased their homemade products (Milo powder and chocolate) in town and explained how they are made. They invited people to taste the product.

Learning Standard 4.2.2: Describe how a cash crop is grown, managed, harvested and processed.

Practical Implementation: Students studied cocoa and practically prepare the cash crop from nursery to field transplanting. They clean and prune the matured trees and pick the ripened cocoa pods from established bearing trees. Ferment and dry the beans in the open sun heat and prepare them homemade milo and chocolate or for exporting in large quantities.

Learning Standard 5.2.2: Identify and describe the main local agricultural activities.

Practical Implementation: Students studied cocoa cash crop and its viability in Kiunga, Western Province. The students prepared the cash crop from nursery to field transplanting. They cleaned and pruned the matured trees, picking the ripened cocoa pods from established bearing trees. Students also fermented the cocoa beans and produced homemade milo and chocolate. They invited the parents and community people to come and taste their homemade milo and chocolate cake. They prepared infographic charts and carried out awareness building in the community to educate people to plant cocoa for income generation and agriculture business development.

Mathematics standards are across Grades 3, 4, and 5 and they correspond to Unit 2 and Learning Strand 1: 3.2.1, 4.2.2, 5.2.1. which aim to teach students how to communicate mathematically by collecting, representing, analysing and evaluating information and how to apply mathematics processes, knowledge and skills in everyday life (PNGDOE 2017a). They will use Quantities and Measurement where they discover and formulate various measures and appropriate units of measurement including length, area, volume, time, weight, money and angle.

Learning Standards 4.2.2: Develop and understanding of the units of measurement and explore area of squares and rectangles such as centimetre squared (cm2), metres squared (m2) and develop formulae to calculate area.

Practical implementation: Students measure the height of the seedling plants at a weekly interval. They survey the planting field and calculate the area to find out how many cocoa trees are needed to plant given the requirement of 5m × 4m per tree. Students harvested the cocoa beans then dried and ground into 300-gram portions.

Learning Standard 3.2.4: Understand the value of various notes and coins as part of a unit and multiples of unit money and solve various money problem situations.

Practical Implementation: Student learned and kept the basic record of the poultry financial records and calculated the profit at the end of the sales of the live chicken. They graphically represent how much money would be made each month, predicted and estimated a twelve-month projection of income and net profit.

8.5.5 Pedagogical Frameworks Used for the Projects

- The STEM-PBL model (Forbes et al. 2021) was adopted to enable the students think about task where students had to unpack a process (e.g., describe the process for making products that can be sold for cash). This framework enabled them to think the process through critical questions across the 5 phases (Ask, Imagine, Plan, Create and Improve).
- Bybee 5E Model

The classroom teaching and learning is done through the application of the Bybee 5E Model (Bybee 2010).

Engagement: Students learn the topic and connect to the real world and value the knowledge acquired as it is useful for life after school and beyond.

Exploration: Students are led to discuss and explore more about ideas and skills required for the topic and its practical approach

Explanation: Students provide their opinion and thoughts on the topic and its contents, and question how it is applicable in real life.
Elaboration: Students ask questions and receive answers form each other, the teacher provides clearer insights to help deepen students understanding on the topic and its project.
Evaluation: Students judge each other's opinion and give feedback verbally. They also evaluate themselves through summative written tests and each day's project phase.

- **The EntreComp Framework**

The Entrepreneurship Competence Framework (Bacigalupo et al. 2016; European Commission 2018) was adopted as the general guiding model for the prototype enterprising activities at scale for learning purposes (see Fig. 8.15). It presents a comprehensive set of skills, attitudes and values for the entrepreneurial mindset required to overcome our current societal challenges. The EntreComp tool analyses three competence areas: Ideas and Opportunities; Resources; and 'Into Action'. Each area presents five competences which are important to develop an entrepreneurial mindset.

Competency 1. Ideas and opportunities: The interdisciplinary curriculum's unit of learning has given an opportunity to discuss and expand classroom lesson to real-world opportunities as the knowledge learned is valuable, workable and sustainable.
Competency 2. Resources: Students and teachers were motivated to mobilise resources and others to undertake the cocoa and chicken project for learning and earning.
Competency 3. Into Action: Students and teachers took the risk to take the initiative and implemented the project as planned. Although there were risks, ambiguities and uncertainties, the projects were a great opportunity for learning.

- **The John Hattie Visible Learning Framework (2014):** Applying the Surface, Deep and Transfer Learning (Modified from Hattie and Yates 2014).

Learning About, *Learning for* and *Learning Through* stages were adopted from the EntreComp Framework (Bacigalupo et al. 2016; European Commission 2018). This combined learning framework below underpins the STEM-PBL for entrepreneurship in school and expansion for self-employment after school (Table 8.8).

8.5.6 Assessment

This is an authentic assessment drawn directly out of the context and needs based on the lived environment for the students. Assessment is carried out throughout the different stages of the STEMpreneur PBL. The assessment is not just done on students, but students are part of the assessment of the STEMpreneur PBL model. The teachers and students also conduct the project itself as a living mechanism. Students wrote a simple strategic plan of how the STEMpreneur project will be implemented. Using a rubric, the teacher guided the students to write and submit their plan for marking. Students also kept journals for each activity throughout the project and submitted them for marking at project completion.

Students' learning in STEM-PBL was measured through written assessment questions in specific subjects. For example, in a science class test, students are asked to explain a situation observed and give a conclusion as per science explanation. On the other hand, the mathematics test has a component for student to calculate quantity and measurement. In this way, we give a numerical value to what student have been learning.

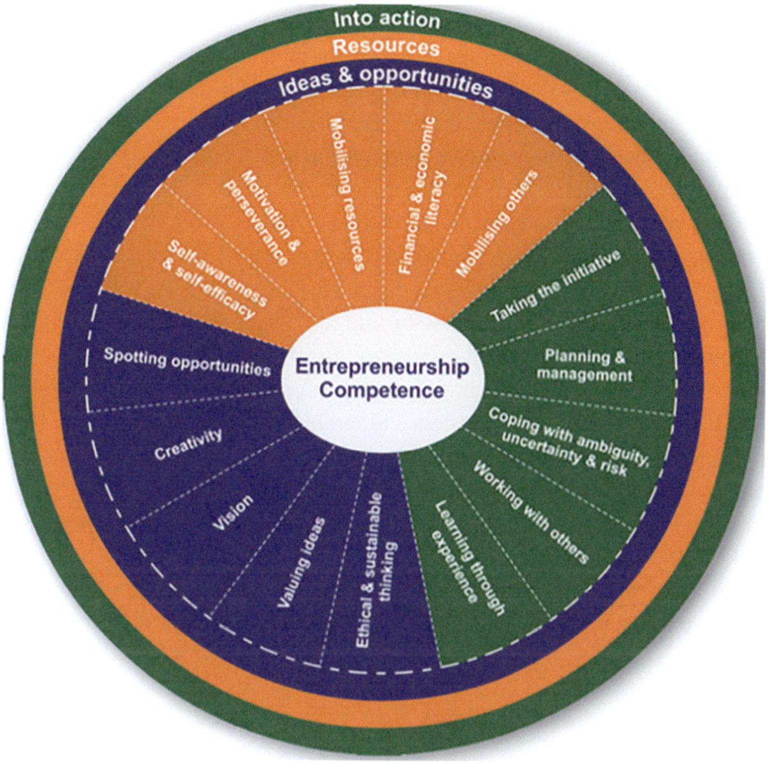

Fig. 8.15 The EntreComp conceptual model (European Commission 2018)

Table 8.8 Visible learning

SURFACE (KNOWING) CONTENT KNOWLEDGE *Direct Teaching of Content* *Learning "About"*	DEEP (RELATING) CONTEXTUAL KNOWLEDGE *Teacher-Student relating to context* *Learning "For"*	TRANSFER (APPLYING) CONSTRUCTIVE KNOWLEDGE *PBL mirroring the future of work* *Learning "Through"*
Curriculum: WALT-K (WALT-K = we are learning to know) Students typically acquire the knowledge set through books, notes and explanation of what is existing based on the curriculum unit of learning Science of plants and animals Mathematics of measurement	*Context: WALT-R* *(WALT-R = we are learning to relate)* Students learn more about a specific project and prepare for implementation as PBL activity. Conceptual learning is connected to a specific real-world project The cocoa and chicken project with their specific parameters	*STEAME PBL: WALT-A* *(WALT-A = we are learning to apply)* Students learn 'through' the practical application of entrepreneurial activity. Such learning requires the development of enhanced reflection skills and relates to practical activities The setting up of chicken coop and raising chicken whilst setting up cocoa nursery and field works

For future projects, there is an aim to contact an industrial partner to observe what the students are doing. For example, inviting a botanist or agriculture scientist who does crafting to cocoa seedlings to assess the crafting process by students.

8.5.7 Challenges Encountered and Overcome

There have been numerous challenges and constraints encountered during the actual implementation of the two projects at the same time.

One of the major challenges faced during the project was almost fifty chicks dying in the first week. We realised that they were exposed to extreme cold, so we installed more light tubes to save them. They also ate sawdust with the chicken feed, so we decided to put grass for them to sleep on. We reported the deaths to the supplier, and he replaced the fifty chicks over time to make up for the loss. Sometimes the shop ran out of stock feed and the chicken suffered with hunger. We found supplementary feed like cassava, coconut and sago until we could purchase new stock feed.

The cocoa seedlings were less vulnerable to damage and were handled with ease. However, one minor challenge is when there is no rain, we manually water them. To avoid insect damaging the leaves, we kept them safe in the green shed house and closely monitored them. These projects were well planned out with a high success rate.

8.5.8 Outcomes and Reflection

Overall, the STEMpreneur project generated 16,100 K (roughly USD 4,500). With this we engaged the school carpenter to refurbish and set benches in the existing library where we placed the eight laptops that we were able to purchase along with one Rachel Technology and other necessities. During the undertakings, we planned for a grand opening and sent invitations to the parents. We held an official opening of our computer lab with RACHEL Technology on 16 September 2021 to celebrate the achievement and also commemorate the PNG 46th Independence Day Celebrations.

With this progress we can now implement what was learned in the subject *Digital Fluency for STEM* by using the laptops and RACHEL Technology to teach students ICT and Scratch Coding activities. These same resources and facilities are also used for teachers' professional development. We had a surplus of 2000 K that was budgeted for the next STEMpreneur projects: fish and vegetable farming to be undertaken in the following term.

In conclusion, creating a STEMpreneurship Culture in the school is a way forward to generating new and better ways of teaching, learning, and earning. This inspires students to build resilience and become future leaders of innovations and entrepreneurship to fulfil the catchphrase—"Take-Back PNG".

8.5.9 Postscript

A positive unexpected outcome of this project has been the school's collaboration with local government and the community. For example, the local MP was impressed by our results and encouraged local people to establish cocoa farms. The MP bought seeds from the school and distributed them to the farmers. We have also improved and expanded our chicken farm and now sell prepared frozen chicken products to the community (K40–60). I find it is truly amazing how a mere science lesson can develop into real-world PBL and have significant and ongoing benefit to the entire community.

My work to promote STEM Education in PNG has continued with the support of AAPNG, Bank South Pacific and our local MP (Member of Parliament). A second STEM school has been opened in Alotau (Milne Bay Province) which also focuses on delivering practical lessons for early learners and primary school age students. My advocacy has extended to delivering professional training for both teachers in school and in university and presenting at national conferences.

I am gladdened to see STEM Education achieving its goal through innovation, entrepreneurship, sustainability and community impact.

8.6 Review of Project-Based Learning in STEM in PNG

Vinesh Chandra and Dann Mallet

The teacher narratives shared in this chapter (8.1–8.4) share similarities with PBL-based STEM Education elsewhere and lessons can be drawn, in turn, for implementation in other settings. However, interesting contrasts are also apparent in the PNG teachers' work. For example, a narrative in this chapter is directly concerned with the financial viability of the school and students' learning experiences are designed to be self-funding.

The PNG approaches to implementing PBL similarly involve some approaches based on problem identification, curriculum-relevant solution generation, and assessment through implementation (see Forbes et al. 2021). We analysed the approaches taken by the PNG teachers using Krajcik and Shin's (2022) six features of project-based learning environments:

(1) Projects start with a driving question that students find meaningful and that creates wonder to sustain engagement and drive learning

The teachers who adopted the PBL approach elected to focus on projects that targeted overcoming problems of importance to the community, broadly defined. Paula Niandros Polume (access to clean water, 8.1), and Levesa Morumo and Abel Mambo (access to electricity, 8.2 & 8.3) centred their students' learning around problems relevant to their local communities, while Elias Wranga (integrated farming, 8.4) (Wranga, et al. 2021) concentrated students' efforts on financially supporting the schools themselves. In each case, a relatively straightforward outcome is identified based on the community problem of interest.

The teachers then follow similar pathways to achieve student learning through a project-based approach. For example, Paula employed the STEM-PBL approach of Forbes et al. (2021), asking about the issue of community water supply, before imagining, planning and creating a solution using the "tippy tap" concept. The driving question in Levesa's activity; how can students create a renewable energy solution for their school? challenged students to think outside the box. Levesa, with a colleague, also adopted the STEM-PBL framework. He began by identifying community electricity supply as a problem before connecting this to the local curriculum and designing appropriate assessment strategies. He finally led his students through the creation of mini hydro-electric plants. Abel also wanted his students to explore renewable and sustainable energy solutions for his school. The driving question for his activity could be stated as follows: How can solar energy be used to solve the problem of unreliable electricity supply at school? Finally, Elias sought to overcome the problem of financing a school, to support students' STEM learning as they built an integrated cocoa and chicken farm to raise funds for school operations. His driving question could be stated as: How can students learn STEM and entrepreneurship simultaneously while addressing school funding shortages and community needs?

Of particular interest, however, were two contrasting characteristics of the work of the teachers in PNG. First, in more developed countries, often, the "context" or "problem" utilised in STEM-PBL must be fabricated in some sense. For example, Bennett (2023) investigated the impact of a Makerspace Learning Environment on student participation and performance in Information Technology classes. He conducted this research in an all-girls school in an Australian city, and his research sample comprised Year Nine students. To develop their knowledge and understanding, the students were presented with a challenge to design and build a model of an articulated bridge. The context was that the bridge which crossed the Meghna River in Bangladesh had been washed away in recent floods. The loss of this critical link impacted the people who lived in the Narsingdi village because it provided the only affordable way to bring food in for the villagers. The students in Bennett's class were challenged to build a model of an articulated bridge that could be made into a full-size replica. In this instance, the students watched videos, gathered relevant information, and developed an empathy for the people whose lives were impacted. Because they were at a distance, it was very different in nature to the immediate problems that the PNG students were addressing.

(2) Students' focus on learning goals that require students to demonstrate mastery on key ideas and practices

In all narratives, it was evident that the teachers aligned the PNG curriculum with the knowledge and skills that their students needed to demonstrate as they engaged in their projects. Different assessment strategies were put in place to develop this understanding. The connections made with the curriculum were very targeted. As per the curriculum's aspirations, local and traditional knowledge were also incorporated in what students did. For example, students demonstrated their science curriculum knowledge (Science unit 9.1–9.4) as they explored locally available materials for the "Tippy Tap" activity. Planning, designing and making the product enabled the students to demonstrate their knowledge and skills in the Design and Technology Unit (9.1–9.4). By applying their measurement skills, students demonstrated their learning in Mathematics as mandated in the curriculum (Mathematics 9.1–9.4). A variety of assessment approaches were evident which enabled the teachers to keep track on the mastery of the identified learning outcomes. Paula's assessment was spread over 10 days, Levesa developed a rubric to assess students' learning across the identified curriculum objectives. His students developed a ranged of conceptual ideas in topics such as work and power, fluids, electricity and electromagnetism. His activity was tied to the Grades 9 and 11 Science, Math, Physics, and Design and Technology curriculum. The prototype created by his students enabled them to learn about some critical, yet difficult concepts such as how DC generators and motors work and how their functionality can be reversed. They also had the opportunity to delve into exploring possibilities for upscaling their models to generate 240 V. Abel was able to connect his solar power bank project to a range of concepts in Year 11 Physics, Mathematics, and Design and Technology. For example, connecting topics in electricity and electronics such as the applications of Wheatstone bridges and potentiometers in classrooms can be challenging.

Elias's project demonstrated some very meaningful connections to PNG's Standards Based Curriculum. The project not only paved the way for developing a STEMpreneurship culture and a business mindset, but it also enabled the students to develop new knowledge about plants and animals and their interrelationships. Elias's project targeted primary school students which demonstrates that these real-world skills can be developed from an early age. His project connected STEM learning using project-based learning (PBL) with the school curriculum to achieve national learning standards. His initiatives were informed by pedagogical frameworks such as Bybee's 5E, STEM-PBL, Visible Learning and EntreComp which guided instruction and targeted outcomes. This innovative project was overarched by STEMpreneurship Culture which embraced "the dual nature of learning and earning simultaneously where teachers can measure their (students) learning effort through multifaceted assessment and income or expenses". Designing classroom activities that promoted the dual nature of learning within the local context informed the driving question.

> (3) Students explore the driving question by participating in practices—processes of problem-solving that are central to expert performance in the discipline. As students explore the driving question, they learn and apply important ideas in the discipline

The teaching and learning sequences were overarched by pedagogical frameworks which effectively targeted the driving questions. The questions were in turn related to a real-world problem or issue which connected with the students' contexts. The need for clean water that was accessible to users foregrounded the activity. Paula outlined how her activity would be sequenced using the STEM-PBL Framework (Forbes et al. 2021). Their students engaged in all aspects of the design process (investigate, plan, create and evaluate) which is a hallmark of how problems and challenges are tackled in technology and engineering subjects. Paula's initiative also showed evidence of a similar design process which was underpinned by five steps—Ask, Imagine, Plan, Create and Evaluate.

The lack of a reliable power supply in his school and community formed the basis of Levesa's driving question. The stimulus questions on "generating electricity" and "mathematics in our community" given to the Grade 9 students were well designed to elicit some thoughtful and meaningful responses from them. The questions created a purposeful environment for hands-on engagement. The Grade 11 students then built on this foundation to connect concepts that were meaningful to their learning in Physics and Mathematics. The evidence of their knowledge and understanding was evident in the generator models they designed and built. Here too, the STEM-PBL Framework (Forbes et al. 2021) was a useful framework to guide their learning. Abel's activity was also driven by the need of a more reliable power supply in PNG. The project has the potential to create a range of learning opportunities for the students and the key areas included: measurement, work power and energy, electrical principles, electronics, geometry, and product design and testing. All these topics connected with the curriculum that could also become a part of an extended investigation. Elias's driving question involved students being led through a customised learning framework building from surface content knowledge to transfer constructive knowledge. One of the cornerstones of Elias's projects was that students learning activities were an integral part of the entrepreneurial initiatives that were led by Elias and his staff at the Cross Country School of Excellence in Kiunga. His students gained conceptual understandings through hands-on activities such as germinating and nurturing cocoa seedlings, raising chicks and monitoring their health, measuring plant height and calculating land area, and producing and marketing homemade Milo and chocolate.

(4) Students, teachers and community members engage in collaborative activities to lead the process of deep learning and to engage in shared knowledge building and to find solutions to the driving question

The activities were either delivered solo or in partnership with a colleague. Levesa worked with Jimmy Tai to involve two different year levels at their school. While delivering lessons in teams or team teaching is not uncommon, especially in integrated contexts, what was less common was collaborating with members of the community. Paula, for example, collaborated with local councillors and leaders so that the Tippy Taps could be installed in their respective villages. In Abels' activity, students would collaborate in groups under their teachers' supervision, along with consultations from the school administration and external partners. Elias's initiative enabled students to work together across grade levels (Kindergarten to Grade 5), often in teams. These collaborative efforts extended beyond students to include teachers, community members, and technical experts. Elias highlighted that his vision led to a "positive unexpected outcome" which set in motion an ongoing collaboration with the local government, Australia Awards in PNG, Bank South Pacific, and the community. The local member of parliament (MP) visited his school and was thoroughly impressed given that Elias was one of the pioneers of growing Cocoa in his area. With this success, the MP encouraged local members of his electorate to establish cocoa farms. The MP bought seeds from the school and distributed them to the farmers. The chicken farm also expanded and was capable of processing and selling frozen chicken. A second STEM school has been opened in a neighbouring province. Elias's comment is noteworthy and one that other teachers around the world can take note; "I find it is truly amazing how a mere science lesson can develop into real-world PBL and having significant and ongoing benefit to the entire community".

(1) *While engaged in the practices of disciplines, students are scaffolded with learning technologies that help them participate in activities normally beyond their ability.* The teachers were based in rural and remote parts of the country where access to learning technologies and connectivity are both significant issues. For these reasons, the use of learning technologies, particularly digital technologies in the activities was minimal as it was only reported by two teachers. Paula's activity was

adopted from models of "Tippy Taps" sourced through the Internet. However, students in her class were also expected to take photos of their projects at strategic junctures so that it could be included in their written reports. Abel's activity was scaffolded by the STEM-PBL framework model with assessment tasks that included journals, reports, and presentations. While not explicitly stated, the activity has significant potential for incorporating technologies. Similarly, the STEM-PBL framework was also very worthwhile in scaffolding students' learning. Digital technologies were used to sketch students' design ideas (see Figs. 8.5 and 8.6) which would draw upon students' knowledge of CAD technologies. Elias's co-authored paper "Teaching science using digital technologies: Some insights from Papua New Guinean schools" was published in a special issue of Teaching Science (Wranga et al. 2021). In this paper, Elias highlights the role that technologies play in implementing scaffolded online research activities. To create opportunities for online activities, Elias was venturing towards the acquisition of RACHEL Pi Technology. This device has the potential of opening new doors for undertaking "offline" research on the internet (Chandra et al. 2016). Among its many capabilities, offline libraries can also be setup which also gives teachers the opportunity to design activities in a more controlled environment. Learning was also scaffolded using frameworks like: EntreComp for entrepreneurial mindsets, Bybee 5E for progressive concept exploration and SMART goals for structured project execution.

(2) *Students create a set of tangible products that address the driving question.* These are shared artefacts, publicly accessible external representations of the class's learning. In each case, the students were able to create a product as a solution that targeted the identified problem. Paula's focus was on setting up an access point for clean water. As mentioned, the Tippy Taps designed and created by the students were installed in local villages where access to clean water was an issue. Tippy Taps were welcomed as an effective solution by the community and, as a consequence, Paula decided to continue with the project with a different cohort of students. Levesa's project delivered a model of a 12 V hydro-electric power generator which was also well received. The project teams believed that, with some modifications, they could make electricity more accessible to schools and communities in rural and remote areas. When implemented, Abel's project would deliver working solar power banks built with recycled batteries and solar panels to power school devices and reduce outages. Elias wanted to make his school and his students financially independent and to use STEM to assure a sustainable future. This would be achieved through selling a range of products., This outreach to the community has gained widespread acceptance and support. As with the study by Rocard et al. (2007) (and many others), the teacher narratives also describe high levels of student interest and motivation with regard to their learning.

8.7 Final Thoughts

The implementation of integrated STEM-PBL learning experiences in the context of schools in Papua New Guinea provided a relatively unique opportunity to inform approaches to STEM Education elsewhere, both in contexts new to STEM Education as well as in locations where STEM Education is more mainstream. In the teacher narratives presented in this chapter, the authors discuss various lessons that were learnt as a result of their journeys that provide guidance for others involved in STEM Education. At an operational level, it is noted that cross-disciplinary collaboration among teachers is vital to creating authentic integrated STEM experiences. Similarly, multiple authors note the importance of genuine collaboration with community members in order to create authentic problem contexts which will motivate student learning and engagement. Caution is advocated when it comes to authentic problem contexts though, because with real-world problems come the real difficulties faced in everyday life—things can really go wrong, like chickens and crops dying, in less controlled learning situations. Nonetheless, the authors demonstrate that successfully addressing community problems through school strengthens relationships and understanding among the community, be it through increasing parents' buy-in to their children's education or through community involvement and support of schools.

References

Amineh RJ, Asl HD (2015) Review of constructivism and social constructivism. J Soc Sci, Lit Lang 1(1):9–6

Andrade H (2000) Using rubrics to promote thinking and learning. Educ LeadShip 57(5):13–18. https://www.ascd.org/el/articles/using-rubrics-to-promote-thinking-and-learning

Asian Development Bank (2008) Medium term development plan III 2018–2022, 2018 (Papua New Guinea). https://lpr.adb.org/resource/medium-term-development-plan-iii-2018-2022-2018-papua-new-guinea

Bacigalupo M, Kampylis P, Punie Y, Van den Brande G (2016) EntreComp: The Entrepreneurship Competence Framework. Publication Office of the European Union, Luxembourg https://doi.org/10.2791/593884

Bennett P (2023) An investigation of the impact of a makerspace learning environment on student participation and learning in information technology classes. Professional Doctorate thesis, Queensland University of Technology. https://eprints.qut.edu.au/241961

Blackley S, Howell J (2015) A STEM narrative: 15 Years in the making. Aust J Teach Educ, 40(7):102–112. https://doi.org/10.14221/ajte.2015v40n7.8

Brophy S, Klein S, Portsmore M, Rogers C (2008) Advancing engineering education in P-12 classrooms. J Eng Educ 97(3):369–387. https://doi.org/10.1002/j.2168-9830.2008.tb00985.x

Bybee RW (2010) Advancing STEM education: A 2020 vision. Technol Eng Teach, 13(6):30–35

Cabrera D, Cabrera L, Powers E, Solin J, Kushner J (2018) Applying systems thinking models of organizational design and change in community operational research. Eur J Oper Res 268(3):932–945. https://doi.org/10.1016/j.ejor.2017.11.006

Camesano TA, Billiar K, Gaudette G, Hoy F, Rolle M (2016) Entrepreneurial mindset in STEM education: Student success. In: Venture Well proceedings of open, the annual conference. National collegiate inventors & innovators alliance, p. 1

Capraro R, Capraro MM, Morgan J (2013) STEM project-based learning: An integrated Science, Technology, Engineering, and Mathematics (STEM) approach. Sense.

Chandra V, Polzin G, Medland R, O'Farell K (2016) The SEE Box: Creating new learning opportunities across STEM disciplines in developing countries. https://eprints.qut.edu.au/101764/14/STEM%2B2016%28Final%29.pdf

Curtis R, Cairns DR, Bolyard J (2023) Design thinking in the middle grades: Transforming mathematics and science learning. Teachers College Press

Department of Information and Communications Technology (2023) Digital transformation agency in Papua New Guinea. https://www.ict.gov.pg

Dewey J (1897/2006) My pedagogic creed. In: Provenzo EF (ed) Critical issues in education: An anthology of readings. Sage, pp 22–31

Doga-Mîrzac M (2017) The promotion of young entrepreneurs for development small and medium business. Valahian J Econ Stud, 8(2):49–56. https://www.researchgate.net/publication/329406802_The_Promotion_of_Young_Entrepreneurs_for_Development_Small_and_Medium_Business

Dugger WE (2010) Evolution of STEM in the United States. In: 6th Biennial international conference on technology education research in Australia, Gold Coast, Queensland, Australia. http://www.iteea.org/Resources/PressRoom/AustraliaPaper.Pdf

Elbaz J, Binkour M, Majdouline I (2013) Innovation and entrepreneurship: An empirical study of Moroccan firms. Paper presented at the AMNET Conference, Algeria. https://www.researchgate.net/publication/272353596_Innovation_and_entrepreneurship_An_empirical_study_of_Moroccan_firms

European Commission. (2018). Employment affairs and social inclusion. http://ec.europa.eu

Forbes A, Chandra V, Pfeiffer L, Sheffield R (2021) STEM education in the primary school: A teacher's toolkit. Cambridge University Press.

Fortus D, Krajcik J, Dershimer RC, Marx RW, Mamlok-Naaman R (2005) Design-based science and real-world problem-solving. Int J Sci Educ 27(7):855–879. https://doi.org/10.1080/09500690500038165

Han S, Yalvac B, Capraro MM, & Capraro RM (2015) In-service teachers' implementation and understanding of STEM project-based learning. Eurasia J Math, Sci Technol Educ, 11(1):63–76. https://doi.org/10.12973/eurasia.2015.1306a

Hattie J, Yates G (2014) Visible learning and the science of how we learn. Routledge

Hunter J (2020) Integrated STEM in Australian public schools: Opening up possibilities for effective teacher professional learning. In: Hunter J, Leping L, Anderson J (eds) Integrated Approaches to STEM Education: An International Perspective. Springer, pp 469–489

Independent State of Papua New Guinea (2010) Papua New Guinea national strategic plan 2010–2050 "2050 Vision". https://png-data.sprep.org/system/files/2011.png.vision.2050.pdf

Independent State of Papua New Guinea (2018) Medium term development plan III (2018–2030)

vol 1. https://png-data.sprep.org/system/files/MTDP-III-Book-1_Final-Proof-Web-compressed.pdf

Kärkkäinen K, Lancrin SV (2013) Sparking innovation in STEM education with technology and collaboration. A case study of the HP catalyst initiative. OECD Education Working Papers, No. 91, OECD. https://doi.org/10.1787/5k480sj9k442-en

Krajcik J, Blumenfeld P (2006) Project-based learning. In: Sawyer RK (ed) The Cambridge handbook of the learning sciences. Cambridge University Press, pp 317–333

Krajcik J, Shin N (2022) Project-based learning. In: Sawyer, R (ed) The Cambridge handbook of the learning sciences (Cambridge Handbooks in Psychology. Cambridge University Press, pp 72–92. https://doi.org/10.1017/9781108888295.006

Lewis T (2005) At the interface of school and work. J Philos Educ 39(3):421–221.https://doi.org/10.1111/j.1467-9752.2005.00446.x

Mbanefo M, Eboka O (2017) Acquisition of innovative and entrepreneurial skills in basic science education for job creation in Nigeria. Sci Educ Int 28(3):207–213. https://doi.org/10.33828/sei.v28.i3.4

Moore T, Stohlmann M, Wang H, Tank K, Glancy A, Roehrig G (2014) Implementation and integration of engineering in K-12 STEM education. In: Purzer S, Strobel J, Cardella M (eds) Engineering in pre-college settings: Synthesizing research, policy, and practices. Purdue University, pp 35–60

Odell MR, Kennedy TJ, Stocks E (2019) The impact of PBL as a STEM school reform model. Interdiscip J Probl-Based Learn 13(2). https://doi.org/10.7771/1541-5015.1846

Papert S, Harel I (1991) Constructionism. Ablex

PNGDOE (Papua New Guinea Department of Education) (2006a) Design and technology subject field lower secondary syllabuses. https://educationpng.gov.pg/Teachers/secondary/lower/syllabus-lower-secondary-design-technology.pdf

PNGDOE (Papua New Guinea Department of Education) (2006b) General mathematics upper secondary syllabus. https://www.education.gov.pg/syllabus/secondarysyllabus/subjectssyllabus/General%20Mathematics%20Grade%2011%20&%2012%20Syllabus.pdf

PNGDOE (Papua New Guinea Department of Education) (2006c) Mathematics Lower Secondary Syllabus. https://educationpng.gov.pg/Teachers/secondary/lower/syllabus-lower-secondary-mathematics.pdf

PNGDOE (Papua New Guinea Department of Education) (2006d) Science lower secondary syllabus. https://educationpng.gov.pg/Teachers/secondary/lower/syllabus-lower-secondary-science.pdf

PNGDOE (Papua New Guinea Department of Education) (2008a) Design and technology upper secondary teacher guide. https://www.educationpng.gov.pg/Teachers/secondary/upper/teachers-guide-upper-secondary-design-technology.pdf

PNGDOE (Papua New Guinea Department of Education) (2008b) Physics upper secondary syllabus. https://www.education.gov.pg/syllabus/secondarysyllabus/subjectssyllabus/Physics%20Grade%2011%20&%2012%20Syllabus.pdf

PNGDOE (Papua New Guinea Department of Education) (2017a) Mathematics syllabus: Primary years 3, 4 & 5. https://www.education.gov.pg/syllabus/primarysyllabus/subjects&syllabus(SBC)/col1juniorprimaryss/jp-mathematics-syllabus.pdf

PNGDOE (Papua New Guinea Department of Education) (2017b) Science syllabus: Primary years 3, 4 & 5. https://www.education.gov.pg/syllabus/primarysyllabus/subjects&syllabus(SBC)/col1juniorprimaryss/jp-science-syllabus.pdf

PNGDOE (Papua New Guinea Department of Education) (2017c) Social Science syllabus: Primary years 3, 4 & 5. https://www.education.gov.pg/syllabus/primarysyllabus/subjects&syllabus(SBC)/col1juniorprimaryss/jp-social-science-syllabus.pdf

PNGDOE (Papua New Guinea Department of Education) (2020) The national education plan 2020–2029: Achieving quality education for all. https://www.education.gov.pg/admin/images/products/b98c46340f30348570b694bc4af6eae5.pdf

Pringle RM, Lord CG, Sheppard TD (2020). Secondary STEM learning. In Johnson C, Mohr-Schroeder MJ, Moore TJ, English LD (eds) Handbook of research on STEM education, Routledge, pp115–127

Rocard M, Cesrmley P, Jorde D, Lenzen D, Walberg-Herniksson H, Hemmo V (2007) Science education now: A renewed pedagogy for the future of Europe. European commission. http://ec.europa.eu/research/science-society/document_library/pdf_06/report-rocard-on-science-education_en.pdf

Talebi K (2015) John Dewey—Philosopher and educational reformer. Eur J Educ Stud 1(1):1–13. https://files.eric.ed.gov/fulltext/ED564712.pdf

Tseng KH, Chang CC, Lou S-J, Chen W-P (2011) Attitudes towards science, technology, engineering and mathematics (STEM) in a project-based learning (PjBL) environment. Int J Technol des Educ 23(1):1–16. https://doi.org/10.1007/s10798-011-9160-x

United Nations (UN) (n.d.) Sustainable development goals. https://www.un.org/sustainabledevelopment/sustainable-development-goals

Weegar MA, Pacis D (2012) A comparison of two theories of learning—Behaviorism and constructivism as applied to face-to-face and online learning. https://www.semanticscholar.org/paper/A-Comparison-of-Two-Theories-of-Learning-and-ivism-Weegar-Pacis/2cb0b3019f04cfbf790d0ed8fa39f603bcfa4f7e

Williams PJ (2019) The principles of teaching and learning in STEM education. AIP Conf Proc 2081:020001–020007. https://doi.org/10.1063/1.5093996

Wranga E, Aku G, Otima F, Chandra V (2021) Teaching science using digital technologies: Some insights from Papua New Guinean schools. Teach Sci 67(3):21–24

World Vision Australia (2023) Tippy taps. https://www.worldvision.com.au/global-issues/work-we-do/water-sanitation/tippy-taps

Vinesh Chandra is an Associate Professor and the STEM Subject Area Coordinator at Queensland University of Technology (QUT). In 2024, he was awarded an Australian Award for University Teaching (AAUT) for his contributions to advancing STEM education in Australia and overseas. He also served as one of the leaders for the PNG STEM Education Teacher In-service Project. For this project, the team was awarded Gold in the QS Wharton Awards and a High Commendation in the IEAA awards. In addition to being co-editor of this book, he authored this chapter and contributed to others. He has also co-authored other books on STEM education, including the award-winning "STEM Education in the Primary School: A Teachers Toolkit" published by Cambridge, and "My STEM Workbook" for primary school students.

Dann Mallet is Professor of STEM Education and Professor of Mathematics at Queensland University of Technology. He is experienced in academic and curriculum team leadership and mentoring, as well as extensive experience as a university Mathematics and STEM educator at all levels. Dann has initiated and led significant international education aid projects in the Indo-Pacific region, and has been recognised through numerous awards and nominations, including, most recently, a Gold in the QS Wharton Awards and High Commendation in the IEAA awards. Aside from his role as co-editor, he has contributed to Chapters 3 and 4. Along with Vinesh Chandra, he co-wrote the Introduction to and Review of Chapter 8 Project-Based Learning.

Paula Niandros Polume was, at time of writing, a Mathematics and ICT teacher at ECOM (Evangelical Church of Manus) Secondary School, Manus Island with almost 10 years of teaching experience. She has since been promoted to Subject Master of Mathematics, pending further advancement to Head of Mathematics. She advocates STEM Education by leading in-services, producing a resource book and supporting funding efforts. She has returned to Queensland University of Technology, through an AAPNG scholarship, to undertake a Masters degree in Education. Her contributions to this book are through Chapter 6 Authentic Learning through recycling waste materials, and Chapter 8 Project-based Learning providing reliable access to water through the construction of Tippy Taps.

Levesa Morumo teaches Biology, Geology, Chemistry and General Science. He now teaches at Wapenamanda Four Square Secondary School in Enga Province. His narrative was based on a project (with Jimmy Tai) at Kwip Dau Secondary School, Mount Hagen. He has contributed to Chapter 8 Project-based Learning with a project to design and build a prototype of a hydro-electric generator to provide their remote school with a reliable supply of power. He blended Project-based Learning with other strands of science with colleagues in the teaching profession.

Abel P. Mambo has been a Mathematics and Physics teacher for over six years. He teaches at Aiyura National School of Excellence in the Eastern Highlands Province. Prior to this, he taught for three years at Kopen Secondary School (now Kopen National School of Excellence). He would like to extend his knowledge by studying for a Masters degree in Physics. Abel contributed to Chapter 8 Project-based Learning with a project of providing electricity using solar bank technology

Elias Wranga has over 17 years of teaching experience. He established the first STEM school in PNG, the Cross Country School of Excellence, in Kiunga in 2015. He continues to run and expand the STEM school to be more entrepreneurial and sustainable. Based on the pilot project, he has developed three additional subjects: Digital Literacy, Financial Literacy and Entrepreneurship Literacy to enrich students learning. He is working actively on a STEM education flagship program called InnovEd Pacific to establish more STEM schools and train more teachers to drive the agenda of innovative and sustainable education across Papua New Guinea and Pacific Island nations. He contributed to Chapter 8 Project-based Learning with a multidimensional integrated farming project focussing on building a School-Based Entrepreneurship Culture for Learning and Earning (STEMpreneurship).

9 Teacher Leadership in STEM

Terri Bourke, Danielle Gordon, Lonah Tombele, Jesse Waison, Anna Semese, Susan Totori, and Pauline Suvi

9.1 What is Teacher Leadership in STEM?

Terri Bourke and Danielle Gordon

While much has been written about leadership and types of leaders including transformational, instructional, authentic and ecological, the concept of leadership remains contestable, undergoing continual reinterpretation, with doubt as to what practices make someone an effective leader (Leithwood 1999). What is clear is that leadership practice is crucial for developing education systems towards sustainable change (Ainscow and Sandill 2010; DeMatthews and Mawhinney 2014; Theoharis 2007; Waldron et al. 2011).

What is also clear is that the vision of a School Principal as the lone heroic leader is rather outdated (Spillane et al. 2001). Instead, what is attracting much more attention as an important part of school leadership, is teachers as leaders (Wenner and Campbell 2017). This attention has been growing globally through the many comprehensive reviews that have taken place (see Nguyen et al. 2020; Pineda-Báez et al. 2020; Wenner and Campbell 2017; York-Barr and Duke 2004). Various scholars (see Curtis 2013; Mangin and Stoelinga 2008; Muijs and Harris 2006) have concluded that teacher leaders, through collaboration and sharing best practices as part of professional learning are important components of educational reform, uniquely positioned to influence positive change in their school contexts. Indeed, some scholars believe that it is only by active involvement of all levels of individuals within an organisation that any change can take hold (Spillane et al. 2001).

However, the concept of teacher leadership has not always been clearly defined in academic literature. Hunzicker (2017), for example, argued that a definition of the concept remains rather elusive. Seminal writers on this topic, York-Barr

T. Bourke (✉) · D. Gordon
School of Education, Queensland University of Technology, Brisbane, Queensland, Australia
e-mail: theresa.bourke@qut.edu.au

T. Bourke · L. Tombele
Kilakila Secondary School, Port Moresby, Papua New Guinea

J. Waison
Kerevat National High School, Kerevat, Papua New Guinea

A. Semese
Markham Valley Secondary School, Morobe Province, Papua New Guinea

S. Totori
Kabiufa Adventist Secondary School, Goroka, Papua New Guinea

P. Suvi
Sogeri National School of Excellence, Sogeri, Papua New Guinea

V. Chandra et al. (eds.), *Integrated STEM Education in Papua New Guinea*, Sustainable Development Goals Series, https://doi.org/10.1007/978-981-95-0828-0_9

and Duke (2004), in their review of the academic literature on this concept, acknowledge different waves of definitions for how teacher leaders have been understood through time. They used the work of Silva et al. (2000) to demonstrate a first wave of teacher leaders who were in formal roles, e.g. coordinators. These teacher leaders were seen as an extension of the administration concerned with the efficient running of an organisation. However, many scholars agree that conceptions of teacher leadership have moved away from formal positions of responsibility or authority (Hunzicker 2017; Spillane 2011; Torrance 2015). A second wave of teacher leaders seemed to focus more on improving instructional practices while a third wave was recognised as leaders both within and beyond a classroom. York-Barr and Duke (2004) elaborated defining teacher leadership as "the process by which teachers, individually or collectively, influence their colleagues, principals, and other members of school communities to improve teaching and learning practices with the aim of increased student learning and achievement" (p. 287). In this section, we adhere to the definition offered in this seminal work. We also agree that teacher leadership is closely aligned with both instructional and distributed leadership theories.

Instructional leadership focuses on the improvement of teaching and learning outcomes through the guidance and support of teachers. This type of leadership involves setting high standards for curriculum and instruction, providing teachers with feedback and professional development opportunities, and monitoring student progress to ensure that all learners are given the best possible education (Óskarsdóttir et al. 2020). This type of leadership has also been termed learning-centred leadership, leadership for learning or curriculum leadership, as a key focus is on developing and coordinating an effective school curriculum (Gumus et al. 2018).

Distributed leadership was introduced by Spillane et al. (2001) as describing instances where school leadership operates as a collaborative organisational culture where problems are solved as a collective, including parents and the community. Jones (2014) noted a direct connection between social capital and distributed leadership as both are based on the core principles of collaborative practice and social connections.

Many writers have followed in the footsteps of the seminal work from York-Barr and Duke (2004). For example, Katzenmeyer and Moller (2009) defined teacher leaders as teachers who "lead within and beyond the classroom; identify with and contribute to a community of teacher learners and leaders; influence others towards improved educational practice; and accept responsibility for achieving the outcomes of that leadership" (p. 17). Wenner and Campbell (2017) conducted another review of the literature in this space. They defined teacher leaders as "teachers who maintain K-12 classroom-based teaching responsibilities, while also taking on leadership responsibilities outside of the classroom" (p. 140). The notion of influencing beyond the classroom is also evident in other studies such as those by Öqvist and Malmström (2016). Margolis (2012) has referred to this type of teacher as the hybrid teacher leader where they maintain their instructional expertise but lead in some capacity outside the classroom, usually above and beyond their regular duties. Collinson (2012) has described these teachers as "informal leaders" modelling learning and innovation, developing relationships and networking opportunities to extend their own learning and influencing others. Angelle and DeHart (2011) claimed that these teachers go above and beyond the call of duty to serve their school and students.

Other writers, rather than looking at teacher leaders as exhibiting a set of behaviours have taken a divergent way of thinking about teacher leadership. For example, Poekert et al. (2016) maintained that teacher leadership is a stance, a way of thinking and being. Smulyan's (2016) ideas are similar referring to teacher leadership as comprised of dispositions, beliefs, attitudes and values about teaching, learning and leadership. He called this a "stance-based view". These ideas appear to build on some of York-Barr and Duke's (2004) notions around trust, being risk takers or seekers of challenge and growth, although these writers do not specifically name these traits as anything other than posing the question—who are teacher leaders?

To analyse the teacher narratives in this section, we will look to see if our teacher leaders exhibit any of these stances, dispositions or beliefs. We will also use York-Barr and Duke's (2004) domains as a robust way to categorise teacher leadership practices. The domains are as follows: school or district curriculum work; participation in school change; coordination and management; professional development of colleagues; parent and community involvement; contributions to the profession; and pre-service teacher education. By doing this, we identify the teachers implementing STEM in PNG as teacher leaders. Work in this area presents as a significant gap in the literature. While there is a growing body of knowledge about teacher leaders generically, Wenner and Campbell (2017) have emphasised the need for content specific teacher leadership research. There are many STEM teacher leadership development models (see Yow et al. 2021). However, these are not based on the development of STEM teacher leaders in less developed countries such as PNG.

Examples of leading by example in STEM Education are presented in this section through four teacher narratives. Each has shown leadership by identifying needs and opportunities and working independently to help others in their professional learning and practice.

> **TEACHER NARRATIVE 9.1**: Lonah Tombele from Kilakila Secondary School explains how, through her leadership, the teachers at her school collaborated to review and streamline their assessment practices to reflect their integrated approach to STEM.
>
> **TEACHER NARRATIVE 9.2**: Jesse Waison from Kerevat National High School identified the need for teacher professional learning in STEM. He took the initiative to design and lead in-service workshops to raise awareness of STEM Education amongst the teachers at his school.
>
> **TEACHER NARRATIVE 9.3**: Anna Semese, from Markham Valley Secondary School in the Markham District, has used the school farm and her discipline of Agriculture as the focus of transdisciplinary learning experiences in STEM and has exercised leadership in her collaboration with other teachers in her school.
>
> **TEACHER NARRATIVE 9.4**: Susan Totori from Kabiufa Adventist Secondary School in Goroka identified gaps in teacher knowledge through personal experience. She led in-service sessions with the teachers in her science department before branching out to supporting teachers throughout the Eastern Province. She used a WhatsApp group to motivate teachers and to build a sense of community.
>
> **TEACHER NARRATIVE 9.5**: Pauline Suvi, from Sogeri National School of Excellence, has focused on empowering and encouraging girls to participate in STEM subjects. She has plans to create a web platform as a complement to the Youngpela Lida Meri (Young Women Leaders) network she founded.

9.2 Teacher Narrative 9.1

Lonah Tombele

9.2.1 Teacher Leadership in Collaborative Assessment in STEM

Kilakila Secondary has undertaken a few integrated STEM projects in the Science and Technology departments. One such project was described in my narrative concerning our "mini" School Conservation Park (see Narrative 7.2), that narrative focused on Design Thinking. This narrative is about how the teachers integrated the assessment component of this project across the eight departments we have in the school. Each department has its own assessment policy, and each subject has its own syllabuses and teachers' guides (PNGDOE 2023). For this project, however, we needed to work between and across subject departments. If STEM is integrated in the classroom, then its assessment also needs to be cohesive and reflect its learning objectives.

9.2.2 Rationale

In the beginning of the 2022 academic year, the government of Papua New Guinea, through the National Department of Education, began promoting project-based learning at all levels of schooling in the country. They stressed the adoption of a Standard-Based Curriculum and the integrated teaching of STEM (Science, Technology, Engineering, and Mathematics). After reading through the information about the reforms in my country, I sat and thought to myself that I could make a change in my school. I decided to collect information from each department by interviewing the subject Heads. They provided me with the units they cover according to the national education school syllabuses at each grade level. This information was gathered and compiled. I then asked about their assessment. They provided everything in detail according to the new PNG SBC (Standard-Based Curriculum) (PNG Education News 2021). My investigations also focused on how the teaching pedagogies they apply in the classroom can be assessed practically, that is, through a hands-on integrated STEM project. Table 9.1 introduces the problem statement for this project, the connections to the curriculum and global goals, the pedagogy I chose and a brief reflection on the project.

9.2.3 Implementation

It was quickly apparent that the teachers had been unknowingly integrating STEM in their lessons for some time. After realising this, I knew that the next step was to lead a school-based collaborative assessment strategy.

I designed and conducted an in-service workshop, according to the school's schedule, on writing programmes and assessment tasks for each topic (see Fig. 9.1). After checking on the teachers' progress, I organised a second in-service which was entirely focused on different types of assessment. We worked on the development of rubrics and how they are best used in assessing students' outcomes.

During these in-service sessions, we discussed how we could combine STEM assessment across subjects by designing standard marking rubrics that would cater for all subjects and their grade levels. This means that, although one assessable task is given, we can use various rubrics depending on the subject and grade level to assess students. For example, if a Grade 9 Science project included the topic, "Earth and Atmosphere" (Grade 9 Science Unit 9), then we check if Social Science has a similar topic like "Climate and People" or Geography may consider "Climate and Weather Patterns". Teachers could access different rubrics to assess different aspects on a common theme. It was a multidisciplinary approach integrating STEM across a range of subject areas.

In PNG, curriculum policies are created by the National Department of Education and distributed to schools while the departmental assessment policies are created within the school. Test and other assessable task schedules are drawn up by the Deputy Principal (Academic). Summative assessment is normally scheduled in Weeks 3, 6, and 9 while project work extends to Week 10 of each term. The Heads of Departments in each discipline and their senior teachers conduct assessment meetings. Their usual agenda items include:

- finding similar themes in each syllabus per department.
- rearranging the units to suit a shared assessment rubric to allow students to be taught the same things in all subjects.
- identifying challenges and strategising how to overcome them.

We developed multiple rubrics for assessing the mini–School Conservation Park as a school-wide project. The learning experiences included a debate, an awareness activity, practical activities and model making (see Narrative 7.2).

Table 9.1 Narrative 9.1 project summary

Problem	Connections		Domain York-Barr and Duke (2004)	Reflection
Implementation of the new nationwide Standard-Based Curriculum and integrated STEM	*Curriculum Connections*		• School or district curriculum work • Participation in school change • Coordination and Management • Professional development of colleagues • Parent and community involvement • Contributions to the profession	A transdisciplinary teacher community of practice was formed supporting collaborative practices for future projects while also promoting efficient assessment design and consistency in assessment practice
	PNG SBC (Standard-Based Curriculum)			
	SDGs (UN, n,d,):			
	4	Quality Education		
	15	Life on Land		
	17	Partnerships for the Goals		

Fig. 9.1 Teacher in-service workshop

9.2.4 Implementation Assessment

We agreed that rubrics are scoring guides used to assess a student's performance. A rubric has three parts: performance criteria, rating scale and indicators. Apart from its use by teachers, it lets students know what is expected and what will be assessed.

We came up with and implemented several standard rubrics that we currently use in our school. The school administration agreed for us to trial the assessment of students using the multi-rubric system that we developed in our meetings (school-based in-service) (see, for example, Fig. 9.2).

The mini-School Conservation Park (see Narrative 7.2), as a whole school project, was

Fig. 9.2 Teacher meeting to share ideas on integrated assessment in STEM

the logical place to start. It involved students in Grade 12 Biology, Grade 11 Business Studies, Grade 11 Geography, Grade 11 Biology, Grade 10 Agriculture and Grade 7 English. The following summarises the curriculum links, the tasks and criteria used for assessment. The tasks were frequently interconnected, for example, the Grade 7 English students were tested on their comprehension of the Year 12 presentations. Students in Grade 11 Biology and Geography collaborated on the construction of a model of the park which addressed the social, cultural, humanistic and technical aspects of creating the park within the school grounds.

Grade 12 Biology

Curriculum	Unit: 12.1—Ecology. Topic 17—Endangered Species, Environment and Conservation. (PNGDOE 2008d)
Topic	Significance of the mini school conservation park
Activity	To raise awareness, students produced posters and charts containing information on the importance of plants and the relationship between animals and human beings, the risk of carbon dioxide in the atmosphere causing ozone depletion, deforestation and clear felling. They addressed conservation practices in looking after the natural environment, developing our park and growing back trees as a way forward to address the global agenda of climate change
Formative assessment	Continuous feedback to students
Summative assessment	Presentation to Grade 7 students. Group work. 5 members/group. 12 groups in total. 4 presenters/day. Assessment completed within 3 days
Criteria	Content organisation, delivery, enthusiasm and audience awareness, collaboration with peers, use of resources, use of supporting details/facts/figures/data, confidence

Grade 11 Biology and Business Studies

Curriculum (Biology)	Unit 11.2—Transport System in Plants (Importance of Photosynthesis) (PNGDOE 2008b)
Curriculum (Business Studies)	Unit 11.1—Planning for the Future 1 (PNGDOE 2008a)
Topic:	Green space in the school is ideal for future STEM projects
Activity	Debate
Formative assessment	Monitoring through preparation
Summative assessment (Biology)	Argue *against* the topic. The opposition talks on the environment conservation and important roles of plants in the environment causing reduction of CO_2 gas in the atmosphere
Summative assessment (Business Studies)	Argue *for* the topic. These students will take the Government side and stress on industrialisation and how development can change the school
Criteria	General knowledge of the topic; Intelligent use of points of order; critical analysis of issue; logic to refute arguments of opponents; securing attention of audience; effective introduction and conclusion; language fluency and clarity; tone of speech; posture, gesture, eye contact; and timing

Grade 11 Geography and Biology

Curriculum	Various
Topic	Project planning
Activity	Model-making
Formative assessment	Continuous feedback on construction of models
Summative assessment	Models of the proposed park were constructed from scrap materials by students. The final design was chosen by peers and teachers
Criteria	Goal of the project; model background; methods used; design thinking; creativity; durability of the model; colours; effort

Grade 10 Agriculture

Curriculum:	Unit 10. 1—Floriculture (PNGDOE 2006)
Topic:	Planting
Activity	Practical
Formative assessment	Feedback and questioning throughout the tree planting activity focused on plants species, soil type, depth of planting holes and their relative distances
Summative assessment	The tree planting will be assessed on spacing (between rows and plants). Students will also use their prior knowledge on soil management in Grade 9 to see what soil is better for which species and so on
Criteria	Theory knowledge demonstration; spacing between rows and plants; soil observation and technique; quality of work; attendance and participation

Grade 7 English

Curriculum	Unit 7.2—Listening Skills (PNGDOE 2007)
Topic	Green space in the school is ideal for future STEM projects
Activity	Awareness and Comprehension
Formative assessment	Practice
Summative assessment	Quiz

9.2.5 Outcomes

The tangible outcome of our collaboration was the development and use of shared rubrics. We were guided by four aims in designing our school-based collaborative assessment strategy.

1. *Collaborative and supportive professional community:* As a group, we encouraged one another to cover lessons according to the term teaching programmes we made at the start of each term. This is when we gather to list the

similar units and topics taught in each department and schedule them so that the concepts are taught simultaneously. Teachers do not teach in isolation; we all progress together.
2. *Efficiency in the assessment process:* The teachers in the school teach multiple grades so they have more than one set of lesson preparation to do each day. They prepare individual lesson plans and make daybook entries for each subject they take. This means the teaching load for a base-level teacher has 32 periods to teach in a week with the day divided into eight periods. Teachers do not have much time to develop customised rubrics to effectively assess their students, so they will often skip such procedures to save time. To improve this situation, we meet and develop rubrics in advance, so we have more time to focus on teaching. After all, what we teach is the same and the rubrics serve the same purpose.
3. *Managing deadlines:* The teachers were given target dates to complete the assessment by the Deputy Principal (Academic). Collaborating with different disciplines assists us by lightening our teaching load and helping us meet our deadlines. We now can submit our plans in a package rather than as individual departments. In the past, we frequently failed to meet the target dates and went beyond the scheduled times, so this is a way forward to improve on our time management.
4. *Consistency in assessment practice:* From experience, many teachers did not develop rubrics. It is challenging for individual teachers and some lack experience in designing and using them. We collaborated to create a consistent file management system to have a repository of assessment rubrics which can be re-used. This is another reason to collaborate as a school community to develop rubrics for all the subjects we teach, creating a common platform for students.

A further outcome was the achievement of SDGs (Fig. 9.3). These were SDG 4 [Quality Education], SDG 15 [Life on Land], and SDG 17 [Partnership for the Goals].

9.2.6 Reflection

An advantage of the multi-rubric approach is that it saves time. Individual teachers do not need to develop their own and can focus on their teaching. Further, it means that everyone in the project keeps up with the others and they have a shared knowledge of what is happening within the school. The disadvantage is in the time taken in developing the rubrics and maintaining a consistency between them. This is only achieved through collaboration and encouraging an environment of shared concern between teaching staff. Having the support of others makes the introduction of integrated STEM projects in the school possible.

9.2.7 Postscript

My experience in this collaborative assessment project has led to further work with teachers. I am now working directly under the deputy principal academic assisting in teacher training and in-service on Standard-Based Curriculum (SBC). I am also currently coordinating STEM, SBC and NIST (National In-Service Training) at Kilakila Secondary School. Further, I am engaged with Don Bosco Schools in the National Capital District training teachers on Integrated STEM in SBC with ongoing training.

9.3 Teacher Narrative 9.2

Jesse Waison

9.3.1 Teacher Leadership in Introducing STEM to a Secondary Environment

STEM Education is a recent concept and teachers in PNG know little or nothing about it. I decided to prepare a series of awareness papers for my colleagues at Kerevat National High School to bring them to an understanding of STEM Education. I will describe the first of several awareness papers I presented at school in-service sessions. I began by defining STEM in the educational

Fig. 9.3 SDGs met through authentic learning using waste materials

context, followed by an introduction to the 21st-Century skills—the 4Cs (critical thinking, creative thinking, collaboration, and communication) (Thornhill-Miller et al. 2023). My work was based on the belief that STEM Education is going to be implemented well only when all stakeholders, especially teachers and school administration officers, understand it and embrace it.

9.3.2 Rationale

In a paper presented for discussion at The Australian National University in 2019, *Assessing the labour market impact of the PNG LNG project and implications for future projects,* Voigt-Graf and Odhuno (2019) argued that there is a general lack of technically skilled workers in PNG. This is a lost opportunity for economic growth. For example, during the construction phase of the PNG LNG project, only 600 of the 5000 licensed welders needed could be found.

How do we close such a gap? We now have the opportunity to address such an issue. Amongst possible pathways, the government is advocating STEM Education in schools. We do not only need to increase the number of welders but also technical personnel for the computing industry as well as engineering and other fields. These technicians must not only be able to do their job well, but they must also be able to continue self-education by building on what they have learnt, to use or develop sophisticated tools that enhance life in and out of their workplace for the benefit of all PNG citizens. These skills must be inculcated in students from their schooling days. STEM Education can help us achieve these. Justifications for STEM Education include:

- Its capacity to respond to a depressed economy;
- Its potential to develop industries and modernise the economy; and,
- It improves the competitiveness of the global economy.

(Ismail 2018)

PNG has more *how* people than *why people.* The *why people* are those that question the reasons behind the working of a system, while the *how people* are those who use a system without being bothered by why the system works. Through an integrated STEM teaching approach, we can be sure to develop the number of *why* people we need.

Implementation of a STEM curriculum will unavoidably suffer if the purpose of STEM is not clearly understood (Portz 2015). While teachers know the content of Science and Mathematics, they still need to learn about Engineering and Technology because these are more practical or applied subjects. Academics may still struggle to understand unless they get out of the books and into the real world of hands-on practical activities. Less-informed school administrators and teachers may think they have Technology covered by having a computer lab for students or that Engineering is covered if students can plan a viable project on paper. The reality is far more challenging.

Teachers, being the implementers of STEM Education, must be well versed with the three types of knowledge central to STEM Education—technological knowledge (TK), pedagogical knowledge (PK) and content knowledge (CK). They also need to be aware that STEM Education always involves the integration of two or more of the STEM disciplines (Lertdechapat and Faikhamta 2021). Table 9.2 provides a summary of my project in terms of problem identification, curriculum and global connections, leadership domain, and a brief reflection on the outcomes.

9.3.3 Implementation

Awareness begins with defining STEM Education and coming to a shared understanding of it in the context of the PNG curriculum. I framed my in-service around three fundamental questions, namely (i) what is STEM, (ii) what is the purpose of STEM, and (iii) what are STEM pedagogies.

9.3.3.1 What Is STEM?

STEM is an acronym for Science, Technology, Engineering and Mathematics. Each of these disciplines can be defined separately below.

- *Science* is the body of knowledge of the natural world systematically organised to make understanding easier. Students need this knowledge to make informed choices.
- *Mathematics* can be thought of as a science of numbers and deals with quantities and their arrangements. It involves logic, patterns, shapes and is abstract.
- *Technology* can be defined as the practical application of scientific and mathematical knowledge. It also includes the equipment and machinery developed by applying science knowledge.
- *Engineering* involves science and technology. It is concerned with designing, constructing, and use of machines, engines and structures.

It is clear these disciplines are linked, and one cannot be successful without another. Integrated STEM takes advantage of this interconnectedness. Technology and Engineering are the more practical parts of STEM and offer students the opportunity of learning through real world activities. Each of the four subjects places emphasis on critical thinking, problem-solving and innovation. STEM Education integrates the four disciplines in a manner that requires students to solve problems using a cross-disciplinary approach. STEM knowledge is critical in solving problems in industry.

9.3.3.2 What Is the Purpose of STEM Education?

Students in the twenty first Century need to have the resilience to cope with rapid technological change. What used to be done by specialists a decade ago can now be done by anyone. For example, sending money between bank accounts with the use of mobile banking is more convenient than standing in a bank queue for hours. Today you can do this anytime, anywhere and at your own convenience.

Therefore, the purpose of STEM is to develop in young people of this generation what literature commonly refer to as the 4Cs—critical thinking, creative thinking, collaboration, and Communication (Scoular et al. 2020; Thornhill-Miller et al. 2023). These skills, as previously noted, are referred to as the twenty first Century skills.

- *Critical thinking* can be defined to be the ability to think clearly and rationally and to understand how actions and ideas are connected. It is evidenced by observation, analysis, interpretation, reflection, evaluation, inference, problem-solving, and decision-making. In STEM, critical thinking is used in developing solutions to problems.
- *Creative thinking* is the ability to offer new or lateral solutions to solve problems. It is a random group of techniques such as brainstorming, mind mapping, experimentation, and asking questions. In STEM, creative thinking is of particular use in situations where resources are limited and need to be re-purposed.
- *Collaboration skills* are the interpersonal skills that people need to work effectively with others. This is evidenced through showing respect for others' views, generosity, effort, managing priorities, meeting expectations, and

Table 9.2 Narrative 9.2 project summary

Problem	Connections		Domain York-Barr and Duke (2004)	Reflection
Disparate and limited knowledge of STEM and STEM Education	*Curriculum Connections*		• School or district curriculum work • Participation in school change • Professional development of colleagues • Contributions to the profession	Delivery of an initial teacher-in-service STEM Education session to garner further support from all stakeholders for an integrated approach moving forward
	PNG SBC (Standard-Based Curriculum)			
	SDGs (UN, n.d.)			
	4	Quality Education		
	8	Decent Work and Economic Growth		
	9	Industry, Innovation and Infrastructure		
	17	Partnerships for the Goals		

adopting a cooperative approach. In STEM, collaboration skills are needed when students are working in teams and each person brings a particular skill or knowledge set to the task.

9.3.3.3 What Are STEM Pedagogies?

Once the purpose of STEM is established, the next question is how are we going to teach to ensure students can apply their problem-solving skills to their learning and to their future workplaces. Pedagogical approaches proven to be highly effective in the implementation of a STEM curriculum are Project-Based Learning (PBL) (see Chap. 8), Project-Based Learning, and Inquiry-Based Learning (Albion 2015). How then are we going to teach to ensure students can apply their problem-solving skills to create the best solutions to a problem in a workplace environment? It has been argued that teachers who possess a highly developed STEM pedagogical content knowledge (STEM PCK) are likely to achieve better results when teaching STEM (Allen et al. 2016). They are able to create an inclusive environment for effective teaching and learning. Such teachers have their programmes centred on learning activities that mimic real-life scenarios. They are able to effectively use a transdisciplinary or multidisciplinary approach to teaching and learning. Here are some approaches that are proven to be very effective in the implementation of a STEM curriculum.

- *Project-Based Learning (PBL):* It is evident in literature that project-based learning is seen as one of the most useful pedagogies in STEM teaching (Forbes et al. 2021). PBL is set up around a challenge or problem with the final solution in mind. In it, students tackle the task from different perspectives and make use of a variety of resources. This approach makes the classroom a place where error is tolerated, and changes are encouraged. The teacher in this scenario is a facilitator, providing expertise in areas of concern. Expertise from the community is engaged where required throughout the project (see Chap. 8).
- *Problem-Based Learning:*
- Problem-based learning and project-based learning use the same acronym (PBL). They are similar, but the key difference is that students are given a real-world problem that they must analyse and evaluate (Hmelo-Silver 2004). Similar to Authentic Learning (see Chap. 6), it involves critical and creative thinking as there usually is no single or clear solution to the problem. This approach fosters creativity, teamwork, and leadership. An example of PBL would be engaging students to create their own plans to address the water drainage problem in the school.
- *Inquiry-Based Learning:* With inquiry-based learning, the role of students in the learning process is emphasised (Pedaste et al. 2015). They are encouraged to ask as many questions

and trial as many novel solutions as possible. Students develop communication skills from this type of learning. Because it is student-led, they have the choice of the type of inquiries or experiments they wish to make. As with Design Thinking (Chap. 7), the teacher in her/his role acts to prompt curiosity and spark reflection.

9.3.4 Outcomes

After consultations with the school leaders, time was made available for a one-hour session (allocated by the school in-service coordinator) for the first paper to be presented. The turnout was excellent with all teachers present.

The general backing for STEM was obvious when the school principal made a commitment to assist in terms of budgetary allocations to support its implementation in the school. It is also evident, after the first and other in-services that followed, that there was a visible change in the way teachers delivered their lessons. Several needs were identified at the end of the initial session, the most outstanding being:

- More in-service on STEM teaching methods.
- Lack of a computer laboratory that has specialised software installed for students' use. The school currently has a controlled internet connection in place (co-funded by the school and PNG-AUS Secondary School Partnership programme). This service was found to be most useful but more specialised software was required.
- Lack of teaching resources to implement STEM Education, particularly hands-on practical activities.
- Time allocation for STEM activities which may interfere with general school schedules.

My commitment to working with my colleagues to reap the rewards of introducing STEM pedagogies works towards achieving the SDGs (Fig. 9.4). These are SDG 4 [Quality Education], SDG 8 [Decent Work and Economic Growth], SDG 9 [industry, Innovation and Infrastructure], and SDG 17 [Partnership for the Goals].

9.3.5 Reflection

STEM Education, though relevant for twenty first Century learners, comes along with some challenges. The main challenge lies in teachers' willingness to accept the invitation to teach a STEM subject and adopt STEM pedagogies. Some may say "as long as we can help a student get selected into a tertiary institution, why bother about STEM?" The purpose of my in-services and awareness papers is to change such perceptions. Following the first introductory sessions at my school, I have focused on twenty first Century skills, STEM PCK—Project-Based Learning, STEM PCK—Problem-Based Learning, STEM PCK—Inquiry-Based Learning, and Assessment in STEM.

The availability of resources, electronic and task-specific devices that students may need to use is limited. Such resources are not easily accessible in most parts of the country. We could be innovative and use local resources around us, but then when students leave school, they will have to deal with technology and tools that is new to them. This fails in satisfying the aim of STEM to give the students the feel of the real world before they get out there.

I believe for the benefit of STEM teachers, the more challenges we face the more we can help our students to develop the twenty first Century skills they need for their futures. STEM is exciting to take up. It has real relevance in the world today and, to be on par with the rest of the world, PNG needs to get involved. Technology is *running* into our country, and we cannot sit back and learn by trial and error anymore. If STEM Education is a means to be up against the tech-giants of the world, we should start with STEM Education. STEM will help to address the different economic issues we have in Papua New Guinea and help our students deal with real-world issues in the future.

Fig. 9.4 SDGs met through teacher professional learning and STEM awareness

9.4 Teacher Narrative 9.3

Anna Semese

9.4.1 Teacher Leadership in STEM through Agriculture

Markham Valley Secondary School in the Markham District has more than 4000 hectares of arable land that is used for agriculture (Fig. 9.5). It is a boarding school which caters for ~ 1100 students. My plan to promote Agriculture at the school through STEM-PBL grew from undertaking the Graduate Certificate in STEM (see Part A). The idea was first implemented in 2021 in collaboration with the Grade 11 Physics teacher, Mr Kaiban Jr. His students planned and created models of a mini hydro, windmill and irrigation system for the agricultural gardens. Our farm provided a real-world context for hands-on learning in Physics.

I believed the idea was worth implementing because integrated STEM based around Agriculture could enrich our students' learning and create something in one subject that is beneficial to another. This means it achieves the SDG 4 (Quality Education). Students in PNG need quality education which, I believe, is not just about making it to tertiary institutions. The character development of the students is also important as it will enable them to appreciate and be successful in the real world.

Developing agricultural education in our school is important at a time of increasing food insecurity and malnutrition in Southeast Asia and Pacific countries (Davila et al. 2021). It contributes to the achievement of SDG 2 (Zero Hunger) and SDG 3 (Good Health and Well-being). It can also contribute to sustainable livelihoods by giving students skills to transfer to possible future occupations and working towards SDG 8 (Decent work and economic growth), SDG 11 [Sustainable Cities and Communities], and SDG 12 [Responsible Consumption and Production]. Overall, it contributes to achieving SDG 15 (Life on land) offering students to learn firsthand about sustainable integrated farming methods. The Sustainability Development Goals these transdisciplinary projects will achieve is summarised in (Fig. 9.6).

On a practical level, the produce grown on our school farm would supplement the quality and reduce the cost of vegetables served in the school mess. Approximately K300 (70 USD approx.) was spent on vegetables every day, a considerable expense for our school.

9.4.2 Rationale

Agriculture is a sustainable way of life. Agricultural education is important because it provides many opportunities for students to grow, develop and apply the knowledge and skills they have learnt (PNGDOE 2006a). Our school has a strong focus on the development of good farming practices in our students. We have cultivated a good land area with Indigenous crops (e.g. sweet potato (kumara), taro, yam, cassava and banana), introduced vegetables and have also expanded our livestock section with different farm animals.

Many of our students' parents do not have a regular income. Hence, they often give animals to the school farm as a form of payment for

Fig. 9.5 Farming at Markham valley secondary school (Photo courtesy of Mr Alex Konehe)

their children's school fees. When parents make these "payments," we ask them to share their Indigenous knowledge of farming with us. In this way, we establish a partnership with the community in educating our students. And we develop new understandings of how traditional knowledge can link with the curriculum. Our school Principal advocates community partnerships, and their importance has been established in the STEM Partnership Forum Final Report (Education Council 2018).

Our approach to agriculture allows students to learn how different farm enterprises depend on each other through their products and processes and the challenges and benefits it has to offer. For example, integrated farming is an agricultural system that synchronises livestock and crop production with the study of soil as the basis of farming (see, e.g., Ouedraogo and Ayantunde 2020).

A transdisciplinary approach will be adopted as instruction moves beyond just blending disciplines but linking concepts and skills through a real-world context. Students will be encouraged to use the knowledge and skills learnt from the curriculum to improve on and connect with Indigenous practices. Project-based Learning will guide students to be creative in how they design and plan activities in integrated STEM disciplines. It will also encourage them to work collaboratively, communicate effectively and think critically.

9.4.3 Implementation

I was keen to apply project-based learning and a multidisciplinary approach to my teaching in Agriculture but realised that this was already my usual practice. My students are always doing practical activities on crop and livestock production, so they were used to developing solutions to given problems.

I needed to identify a real problem that the school farm is facing which could be solved or minimised through STEM. I began a conversation with a keen young Physics teacher, Mr Kaiban Jr, who was interested in the Graduate Certificate course in STEM Education that I was undertaking. I shared insights on STEM, PBL, twenty first Century skills and writing rubrics. After some time of mentoring him, I challenged him with the problem on reliable irrigation in the school gardens. What he started coming up with not only

Fig. 9.6 SDGs met through teacher collaboration and multidisciplinary STEM integration

Fig. 9.7 Student prototype of windmill to improve irrigation in the school farm

fascinated me but gave me hope of implementing my plan of transdisciplinary STEM to support my teaching in Agriculture.

The first project was to improve the irrigation of the crops at the farm. The students built a prototype windmill (Fig. 9.7) which supported the Physics syllabus (Unit 11.4: Work, Power and Energy (PNGDOE 2008c).

The next project was a sprinkler system constructed from empty bleach containers (Fig. 9.8). This supported student learning in Physics (Unit 12.1: Fluids) (PNGDOE 2008c). By mastering the construction of the sprinkler's swing arms, students demonstrated their understanding the behaviour of fluids.

The windmill was designed to pump water while the sprinkler irrigation system was to water the agriculture gardens. Problem solved! This I saw was a great start in the implementation of my plan. Physics students were given the opportunity to use their knowledge to create products in another context.

Since a transdisciplinary approach will be taken in Project-based Learning, students will be encouraged to integrate knowledge and skills from other STEM disciplines to work on the project. Other STEM discipline subject teachers will be encouraged to work collaboratively with the teachers of Agriculture to provide a context for them to achieve their learning outcomes. Table 9.3 presents a summary of my project.

Fig. 9.8 Student prototype for sprinkler system built from waste materials

Table 9.3 Narrative 9.3 project summary

Problem	Connections		Domain York-Barr and Duke (2004)	Reflection
Making the school farm the context for Integrated STEM projects	*Curriculum Connections*		• School or district curriculum work • Participation in school change • Coordination and Management • Coordination and Management • Professional development of colleagues • Parent and community involvement • Contributions to the profession	Models created from the initial project provide evidence to action a scaled version
	PNG SBC (Standard Based Curriculum)			
	SDGs (UN, n,d,)			
	2	Zero Hunger		
	3	Good Health and Well-Being		
	4	Quality Education		
	8	Decent Work and Economic Growth		
	11	Sustainable Cities and Communities		
	12	Responsible Consumption and Production		
	15	Life on Land		

9.4.4 Outcomes

The collaboration that I needed for the implementation of my idea was a success. When the administrators were invited to the Grade 12 Physics model presentations, they pledged their support for making the models a reality in our school gardens. It was an additional bonus with our students also participating in the Morobe Educational fair in the areas of Agriculture, Science and Technology.

9.4.5 Reflection

The challenge is now to get other disciplines involved in my plan. The transition to the

Standard-Based Curriculum (PNG Education News 2021) has inadvertently brought this closer to reality. I was invited to present an in-service to raise awareness on PBL and STEM. In the presentation, I outlined how the school could be successful in implementing the SBC through a transdisciplinary approach based around Agriculture and the school farm. It was successful and the school administrators are now embarking on identifying problems and tabling major projects that Heads of Departments in the school will take on board according to the revised syllabuses.

The school is now looking forward to developing and promoting transdisciplinary projects in the area of Agriculture. I am looking forward to working collaboratively with all STEM disciplines to schedule joint projects. It is an exciting opportunity to engage our students and make broader use of the skills they learn in separate disciplines.

9.5 Teacher Narrative 9.4

Susan Totori

9.5.1 Teacher Leadership in distributed teacher professional learning in STEM

In the context of Papua New Guinea, the notion of Science, Technology, Engineering and Mathematics (STEM) Education is fairly new to the majority of people working in the teaching field. I came to a personal understanding of the need for teacher development when I proposed a strategic plan on the theme "Integrating STEM in Water Quality Analysis" for my school and local community who also access the water supply. I was immediately hindered by teachers' limited knowledge and perceptions about STEM Education and their reluctance to adopt an integrative STEM-PBL approach in their teachings.

In order to successfully implement the project, I had planned, I realised my teachers needed a fair knowledge and understanding of STEM Education. Because of this, the initial step I took was to plan and carry out teacher professional learning for teachers within my school. I started with the science department through departmental in-services and then, with the consent of the school administration, to the school at large through school-based staff development programmes. Reaching out to other schools within my province was the next stage of my plan. I liaised with the Provincial Education authorities to conduct STEM training in the province. This narrative gives an account of the approaches undertaken in driving the concept of STEM Education to other teachers in Eastern Highlands Province of Papua New Guinea. Since beginning this programme, I have continued to monitor and evaluate the progress of STEM implementation in schools.

9.5.2 Rationale

The Government of Papua New Guinea is now strongly embarking on quality education with the introduction of STEM Education in the newly reformed Standards-Based Curriculum (SBC) (PNG Education News 2021). However, the transition and implementation has been sluggish because the concept of STEM is novel to most educators and the appropriate knowledge and skills about STEM are very much lacking. Therefore, teacher training in STEM and STEM Education is imperative to ensure a smooth transition and implementation of this concept in Eastern Highlands schools and PNG as a whole.

Teamwork in any STEM integrative approach is equally important for the continuity and sustainability of STEM Education. Teachers have to be the centre of attention and must be frequently upskilled with relevant STEM knowledge and skills through efficient teacher professional development programmes for successful implementation of STEM Education. This will positively enhance capacity building in teachers, making them "STEM literate" and confident in collaborating with others to achieve desired learning outcomes.

I shared a grade with another colleague, and I soon came to realise we needed to be on "the same page" to achieve genuine STEM integration. This would not happen if there was little uniformity in our pedagogical approaches. A positive disposition and shared mindset are essential.

Teachers' fluency and flexibility in integrating digital technology in their teaching is also a potential barrier to effective implementation of STEM in PNG. For many teachers in schools, the use of digital technologies often goes hand in hand with fear, apprehension, and uncertainty (Forbes et al. 2021). This was very evident in my school.

Furthermore, at the secondary school level, teachers are more specialised in the subjects they teach, meaning they often have truly little knowledge of STEM disciplines outside of their own. Lack of expertise in each area also can be problematic for implementation. Different teachers have more in-depth knowledge of each field (Chalmers et al. 2017). Therefore, frequent teacher professional development programmes will certainly enable teachers to adapt to a transdisciplinary pedagogical approach and be able to engage meaningfully in team teaching.

Encouraging teachers to support each other to become confident in STEM pedagogies will help towards the achievement of the SDGs (Fig. 9.9). The prime goal is the SDG 4 [Quality Education] in its advocacy of STEM Education. It also supports the achievement of SDG 10 [Reduce Inequalities] in that students, irrespective of which school they attend, will have comparable opportunities to engage in meaningful STEM activities. The community of practice that is being built and sustained through social media is related to SDG 16 [Peace, justice and strong institutions].

9.5.3 Implementation

During the Graduate Certificate course, we were grouped into our respective provinces. For us, as the team from the Eastern Highlands Province, based on our context, we came up with a plan to come together as STEM educators to carry out awareness programmes on STEM Education to all teachers (especially at the secondary schools) in our province. Upon completion of the course, we returned to our schools determined to execute this plan.

Back in our individual schools, we encountered many challenges as we tried to embrace and integrate our acquired knowledge and skills in teaching and learning. For me, being the Head of Science in my school, I had an advantage in influencing other science teachers about the STEM agenda through on-going formal and informal department professional development (Fig. 9.10).

After several in-services conducted with my Science teachers, I took the next step to approach my school Principal. He was convinced and gave his approval with attached funding for school-based training for all teachers at the school. I designed and conducted a two-day in-service on STEM-PBL and assessment.

This was followed by a consultation with the Eastern Highlands Provincial Education Authorities. The Provincial Education Director responded positively to the idea of implementing STEM. He supported the idea of a team of STEM teachers in the province running an awareness presentation. I played a key role in organising and leading the presentation which turned out to be remarkably successful. The Provincial Education Director, School Inspectors and the District Education Services Managers were totally convinced to embrace STEM Education in the Eastern Highlands Province of PNG.

Furthermore, I was able to conduct STEM training for three Adventist high schools and five primary schools in the Eastern Highlands province through a two-day professional development organised by the Director of Education for Adventist Agency Schools. The highlight of the in-service was how effortlessly primary school teachers were able to grasp the idea of integration. I believe that this is because they are general teachers unlike the specialised subject teachers at the Secondary School level. They are more used to combining different disciplines to create learning experiences for their students. Table 9.4 presents a summary of the professional development programme I have designed and delivered.

Fig. 9.9 SDGs met through teacher professional learning and communities of practice

Fig. 9.10 Conducting department in-service for science teachers

9.5.4 Outcomes

As a result of the teacher training I have conducted in my school and other Adventist primary and high schools in the province, teachers are now embracing the STEM agenda and are implementing STEM-PBL activities in their lessons. It is indeed fascinating to see how STEM is gradually gaining momentum in the schools whose teachers attended training.

Science teachers in my school are remarkable in adapting and implementing STEM in their teaching and learning. However, not all other subject teachers are implementing an integrated STEM approach which is a great challenge for me as a STEM advocate. To address this issue, I have created a WhatsApp group for the teachers who attend the training I conduct. I see WhatsApp as a convenient platform for teachers to interact and share their experiences of STEM implementation. This initiative is progressing well, and teachers have been uploading their students' work. How they reflect and evaluate each other's efforts, provide useful feedback and make recommendations for improvement is captivating.

Fig. 9.11 Conducting in-service STEM workshops

Fig. 9.12 SDGs met through leading gender equity in STEM

Table 9.4 Narrative 9.4 project summary

Problem	Connections		Domain York-Barr and Duke (2004)	Reflection
Delayed transition and implementation to new Standard-Based Curriculum due to novelty of STEM and STEM Education by teachers	*Curriculum Connections*		• School or district curriculum work • Participation in school change • Coordination and Management • Professional development of colleagues • Parent and community involvement • Contributions to the profession	Initial support from the school principal allowed for an in-service training focused on STEM-PBL and assessments leading to a broader Eastern Highlands Provincial Education Authority support towards embracing STEM Education through two days of targeted professional development
	PNG SBC (Standard-Based Curriculum)			
	SDGs (UN, n.d.)			
	4	Quality Education		
	10	Reduced Inequalities		
	16	Peace, Justice and Strong Institutions		

Further to this, I am also working with teachers who have shown interest in integrating STEM in their teaching, especially in writing STEM-PBL tasks and developing assessment criteria. This has emerged as a genuine challenge faced by most teachers as they try to integrate STEM into their classroom.

The slow pace of STEM understanding and implementation by teachers signifies the need for more teacher training. Many teachers seemed more comfortable with the pedagogical approaches they are familiar with and not very open to changes. A perceived lack of digital literacy and uncertainty in integrating technology in teaching and learning are obstacles in effective implementation of STEM in schools. Despite these challenges, significant progress has been made in the implementation of STEM Education.

Ultimately, the STEM Graduates in my province, as well as myself, have been recognised as Trainers of STEM by the Eastern Highlands Provincial Division of Education. Recommendations have been made by the authorities for us to conduct intensive one-week training programmes for all secondary and high school teachers in the province. Such training will be conducted frequently so that teachers are very well acquainted with the concept of STEM Education and become confident in integrating it in their teaching and learning.

Finally, monitoring and evaluation is underway. I am using the WhatsApp platform to monitor how the schools are progressing with STEM integration. Phone and verbal communications are used in following up with the teachers concerned. The next step is to physically go to each school and interview the teachers on the ground to get their feedback. The information from social media, phone calls, conversations, and interviews are used to inform further teacher-training sessions.

9.5.5 Reflection

My school is now viewed as an exemplary STEM school in the Eastern Highlands Province because of the work I am doing as an advocate for STEM. We are planning for a Science Fair for the school. The Adventist Schools who are progressing with STEM implementation will be invited to showcase their work in STEM together with our students. Neighbouring schools will be part of the audience. We are anticipating that it will be both educational and engaging for those who attend. It is purposely designed to capture and impact the interest of teachers and students in the province about the importance of STEM Education.

It is important for teachers beginning to implement STEM to see what can be produced through adopting STEM pedagogies. My school has recently purchased a Canon video camera following a request by the STEM teachers. We have started capturing STEM-PBL activities to share with others. Our next step is to ask the school to create a STEM website so we can upload students' activities for other teachers to access and use our ideas in their teaching.

The challenges encountered in getting teachers to better understand STEM and implement it effectively in their classroom are overwhelming in our schools. Some teachers are more comfortable with their pedagogical approaches and are reluctant to embrace new concepts. They offer excuses to disguise that they lack the relevant STEM knowledge and skills. Dedicated effort and time are paramount in building the capacity of teachers.

Personally, I am determined to pursue further study in STEM Education so that I can be fully equipped and qualified to train the pre-service and In-service Mathematics and Science teachers at Teacher Training institutions. I believe we can effectively address the STEM agenda in PNG schools by encouraging our pre-service teachers to graduate with STEM knowledge and skills and the confidence to implement integrated STEM Education.

9.5.6 Postscript

My focus has remained on STEM Education and PNG's Standards-Based Curriculum (SBC). To this end, I have continued conducting training for in-service teachers in Port Moresby and the

Eastern Province and for pre-service teachers and academics at the University of Goroka (see Fig. 9.1). My experience has also been sought in the design of postgraduate courses in STEM. I have expanded my own teaching to include targeted excursions, such as taking Year 12 Biology students to the Institute of Medical Research (IMR) in Goroka.

Through these experiences, I have refined a list of seven aims for teacher professional development in STEM which may guide others. These are:

1. To develop an understanding of STEM, STEM Education and the pedagogical approaches required to develop twenty Century Skills in students.
2. To develop an understanding of STEM integration within the published curriculum.
3. To reflect on the similarities and differences between current practices and integrated STEM Education.
4. To develop and understand the importance of STEM Education in our contemporary society of the twenty first Century.
5. To develop an understanding of STEM-PBL and the importance of developing appropriate assessment criteria (rubrics) to assess students.
6. To encourage teachers to offer more practical oriented lessons rather than teaching theory. (Which is the main current practice in Schools in PNG).
7. To develop teachers' understanding of STEM Industrial Partnerships and organising School Science fairs to enhance students' motivation in Science and Technology related courses and careers.

I remain committed to my advocacy of STEM Education and it is rewarding to see its progress. The WhatsApp group I established now includes all participating schools. Teachers are encouraged to share ideas and post pictures and videos of their STEM activities. Very soon, every school taking part in our in-service workshops will have uploaded a video. Some of these projects will be shown during their Science Open days (Science Fair).

Finally, I am encouraged to find that many PNG teachers are well versed now with STEM pedagogies and are prepared to implement STEM integration in their teaching and learning. Primary Teachers easily grasped the idea of STEM integration in Teaching and Learning more easily than the Secondary School Science Teachers. I think it's simply because they are generalised teachers who teach all subjects, unlike the Secondary teachers specialising in only one or two subjects. Mostly, I can see that teachers are supporting each other in developing and enhancing their STEM pedagogy and supporting student learning.

9.6 Teacher Narrative 9.5

Pauline Suvi

9.6.1 Teacher Leadership in Improving Female Participation in STEM Through a Digital Platform

> Countries with more gender equality have better economic growth. Companies with more women leaders perform better. Peace agreements that include women are more durable. Parliaments with more women enact more legislation on key social issues such as health, education, anti-discrimination and child support. The evidence is clear: equality for women means progress for all.
>
> Ban Ki-moon (2014)

Ban Ki-moon, as the Secretary-General of the United Nations from 2007 to 2016, clearly understood the importance of empowering women in economics, business, politics and government. In PNG, efforts are being made to build female capacity in leadership in education. A critical example of this is *Ol Lida Meri* (All Women Lead), a professional learning course launched in 2021, designed to bring women together to build leadership skills and confidence (PASS, n.d.). The *Ol Lida Meri* programme in 2022 (OLM2), which I co-facilitated, had three goals. These were:

1. Educators and students build leadership skills and use student-centred and project-based

learning approaches to improve educational outcomes.
2. Educators in nominated secondary schools deliver an enhanced STEM curriculum.
3. Positive understanding and relationships between PNG and Australia.

To extend the work of *Ol Lida Meri* to girls and young women currently in secondary school, I have established a new network, *Ol Yangpela Lida Meri*. Adding "yangpela" to the group's name highlights my focus on youth while acknowledging its connection to the existing group. It is currently implemented in all PNG Schools of Excellence and four selected secondary schools in Papua New Guinea.

My particular aim is to address the gender inequality of girls' participation in STEM Education and to prepare girls for a STEM-based job market. The number of girls taking up STEM Education in Papua New Guinea is far less than males. The reasons can be many but the prominent one is that girls lack confidence. They think that they are not good enough compared to boys (Chesky and Goldstein 2018). They have low self-esteem because there is little support and few opportunities to encourage their participation in STEM. According to Prinsleyet al. (2016), if girls are interested, engaged, and motivated to take STEM subjects, they will perform as well as the boys.

Ol Yangpela Lida Meri's vision is to empower young women, especially from rural communities, to develop the capabilities, mindset and confidence to become the next generation of leaders in our country. It is currently an ongoing school-based initiative that engages female students and is planned and refined annually. Female students come together for leadership activities and an off-campus retreat hosted by a female role model or mentor. It is being developed through a PNGAus Secondary School Partnership Activity.

In short, I would like to leverage what I have already started by building a website that can serve as a learning bank and connection hub to encourage and support girls in STEM within and beyond its current school setting. The *Ol Yangpela Lida Meri website* will be an ongoing school-based initiative that will engage female teachers and students to participate, provide feedback and use as a learning resource.

The online platform will address SDGs (Fig. 9.2). These are SDG 4 [Quality Education], SDG 5 Gender Equality], and SDG 8 [Decent Work and Economic Growth].

It also aligns with one of the pillars of PNG's Vision 2050 which addresses Human Capital Development, Gender, Youth and People Empowerment (Independent State of Papua New Guinea 2010).

9.6.2 Rationale

When we consider the bigger picture, women bring a unique experience and perspective to STEM industries (Ki-Moon 2014; Milgram 2011). While women contribute equally with men to the economy of the country, fewer do so through STEM-related fields.

I believe that fewer girls in Papua New Guinea are taking STEM because they subconsciously think that STEM-related jobs are exclusively for males, and they grow up in a society which promotes this stereotype. For example, Forbes et al. (2021) observed that it is common for the image of a male wearing a hard hat to appear when searching for an image of an engineer.

The Girls in STEM Toolkit (Education Services Australia, n.d.) provides positive images of girls involved in STEM. It is an example of what I could do in creating an effective website and provides a model of what I want to do in Papua New Guinea to address this gender gap in STEM.

9.6.3 Implementation

Ol Yangpela Lida Meri will provide a perfect platform to implement my plan. I have used design thinking to plan my project and it will become the framework that I will use to build the project.

- *Empathise.* I understand that fewer females than males are taking STEM classes, and this will limit their future prospects in STEM-related study and employment. Cultural beliefs and gender stereotypes may support the unconscious bias that STEM is too difficult or inappropriate for girls.
- *Define.* To better understand the issue, I shall:
- read the available research on girls' access to and participation in STEM Education.
- identify benefits and challenges of girls in STEM Education and how these relate to PNG.
- consult with other STEM teachers on the experience of girls taking (or refusing) STEM in their schools.
- conduct surveys to better understand the issue. I will determine how many girls participate in STEM classes in Sogeri and other schools in the Yangpela Lida Meri network and I will identify the different online platforms available to girls.
- *Ideate.* The questions I need to address are:
  - How might I make STEM accessible to all girls and encourage them to be interested in STEM so that they can be the next generation of leaders in Papua New Guinea?
  - How can I make STEM accessible and empower female students to take STEM?
  - How can I develop their twenty first Century skills, mindset and confidence as emerging leaders in Papua New Guinea?
  - What are the benefits and constraints on reaching PNG girls through an online platform?
- *Prototype.* The prototype will be developed following:
- collaboratively brainstorming the pages of the website using a mind map.
- creating a design brief.
- consulting female students through an online questionnaire about the design to ensure it is user friendly and easily accessible.
- consulting an IT specialist to advise on the website and the hosting site.
- seeking advice on Policy and Child safety. Check compliance with SEAH (sexual exploitation, abuse and harassment) guidelines.
- A first draft will be made available for review and critique by an advisory committee.
- *Test.* Testing of the website will include:
- launching the website to the school allowing access to teachers to test.
- allowing female students and teachers to access the site to review.
- seeking comments and recommendations.
- comparing outcomes against design brief. Check that the website is fit for purpose.
- compile and consider recommendation and comments.
- making changes as necessary.
- re-testing depending on recommendations.
- repeating the testing process with a second draft using a limited audience of all teachers both inside and outside the school.
- *Evaluate.* Evaluation will include:
- finalising the website.
- launching to other schools.
- monitoring the site's use and impact.

The development of the website will be voluntarily supported by colleagues. I have established multiple advisory committees. The first is an intra-school committee made up of four female teachers with strong backgrounds in technology. The second is an inter-school committee with teachers representing different provinces and expressing an interest in girls' education. Potential members of this committee will represent the Autonomous Region of Bougainville, the Eastern Highlands, the Western Highlands, and the East Sepik Province. Further I have received offers of international support through advocates of *Ol Lida Meri* in Australia and academics at the Queensland University of Technology.

Building and maintaining the website requires physical resources. The working space needed already exists and the technical equipment needed is available. I will need access to a connected laptop or desktop with videoconferencing and graphics software, a phone (to contact regional advisors), a digital camera, a camera stand, lights, microphone and a green screen. Table 9.5 presents a summary of my project.

Table 9.5 Narrative 9.5 project summary

Problem	Connections		Domain York-Barr and Duke (2004)	Reflection
Gender inequality of girls participating in STEM Education	*Curriculum Connections*		• School or district curriculum work • Participation in school change • Coordination and management • Parent and community involvement • Contributions to the profession	The extension of the existing girl's leadership programme, *Ol Yangpela Lida Meri*, to include supporting STEM Education with volunteer teachers to establish advisory committees across multiple regions
	PNG SBC (Standard-Based Curriculum)			
	SD (UN, n.d.)			
	4	Quality Education		
	5	Gender Equality		
	8	Decent Work and Economic Growth		
	17	Partnerships for the Goals		

9.6.4 Outcomes

The intended outcomes of this project are that, through a purpose-built website, I hope to be able to:

- engage with female students in PNG schools to promote a mindset and skills conducive to becoming active participants and leaders in STEM.
- establish support networks for high-performing female students now (as a foundation) and into the future as they navigate the challenges of diverse pathways.
- support female teachers to become active role models for girls and young women in the adoption of STEM subjects and careers.
- establish an off-campus community involving female leaders of PNG in STEM Education, government, and industry.
- encourage girls and young women to know their own strengths and work together to support equity, access, and inclusion.
- support teachers of STEM to reach out to girls in their schools and communities through the uploading of targeted resources and exemplars which the integration of STEM subjects.
- keep girls and young women informed of study opportunities and scholarships in STEM.
- assist girls and young women more generally to know and promote their legal rights, for example, Child Protection, Protection from Sexual Exploitation, Abuse and Sexual Harassment (PSEAH), and Cyber-Awareness (Femili PNG, n.d.).

An important feature of the website is that it is interesting and engaging. To achieve this, I will experiment with including links to hands-on activities which can be done individually or collaboratively:

- Maker Space (see Blackley et al. 2018; Sheffield et al. 2017).
- Tok-Stori (see Sanga et al. 2021) in collaboration with the SEE (Share, Engage, Educate) Digital Library Project (seelibrary.org).

The content and features of the website will change over time in response to users' comments and suggestions. It will also respond to changes in curriculum demands or syllabus updates.

9.6.5 Reflection

In my 20 years of teaching, I have noticed that young men often bring into the classroom the cultural imprimatur that they are leaders and have more rights than females. They do not share the space, nor do they give girls the opportunity to voice their opinions. This means that most young women opt to remain silent during discussions. Through this project, I hope that girls find their comfort zone to have access to positive and useful information, to be active and to interact with others.

On reflection, I think that the most important considerations for my website will be to communicate my ideas clearly and to keep the potential impact and benefits for the school community uppermost in my mind when making decisions. My criteria for success for the site will be that it is:

- *user friendly*—providing easy to follow instructions and navigation, be localised so that girls can view it from their world, be global so that girls can experience broader perspectives.
- *current*—containing relevant and current STEM resources; the blog will be updated monthly.
- *motivational*—including stories of STEM role models.
- *active*—providing links to activities and projects to be completed individually or collaboratively.
- *interactive*—maintaining a regulated platform to communicate through feedback, comments, zoom, and webinar.
- *safe*—complying with child protection and SEAH guidelines.
- *accessible*—allowing access both on- or offline at school or at home.
- *inclusive*—giving girls the freedom to learn without any bias. Girls will be in their comfort zone to get access to information.

Above all, I want to encourage girls to find out more information and perhaps build their interest IN STEM. Motivation in STEM comes before learning can take place. Girls have the potential but just need a spark to get them started. There are courageous women in Papua New Guinea who dare to venture into the STEM fields that are traditionally regarded as male orientated. Their stories are worth telling. The promotion of integrated STEM Education is a part of a larger story which develops the capabilities of girls and young women so that they can lead in the classroom, school, and communities. STEM encourages STEM thinking capabilities which is fundamental for leadership in the twenty first Century.

9.6.6 Postscript

This project is still in its initial stage (registration and policies), but I am proud to say it is a work in progress. My next step is to register the website under the Investment Promotion Authority and Department of Community Development and Religion, so it is currently offline. In addition, I am working on registering the website domain and the layout of the website. Hosting will cost approximately K1000 (~250 USD$). It is hoped that financial assistance will come through access to project funds within the school. Other grants such as the Alumni Association and grants from Australian Awards (AAPNG) might also be a source of funds. I am committed to completing this project and establish a vibrant digital platform to support girls and young women in increasing their skills and confidence in STEM. I intend to make this platform the basis for further tertiary study, as I am currently applying for a scholarship under the Australian Awards.

9.7 Review of Teacher Leadership in STEM in PNG

Danielle Gordon and Terri Bourke

We begin by reiterating the six domains of teacher leadership practice we will use in this review: school or district curriculum work; participation in school change; coordination and management; professional development of colleagues; parent and community involvement; and contributions to the profession (York-Barr and Duke 2004). These domains provide a useful framework for understanding "what teacher leaders do" (York-Barr and Duke 2004, p.266). Additionally, we examine stances, dispositions or beliefs about "who these teacher leaders are" and how they have become leaders. Our teacher leader narratives are from Lonah Tombele (9.1), Jesse Waison (9.2), Anna Semese (9.3), Susan Totori (9.4), and Pauline Suvi

(9.5) and demonstrate how STEM practices were enacted both in their local school contexts and beyond.

9.7.1 School or District Curriculum Work

For each of these teachers, the catalyst towards becoming a teacher leader was the introduction of a new standards-based curriculum initiated by the government of PNG and its promotion, through the National Department of Education, of STEM Education. These teachers understood that to effect sustainable change, they had to look beyond their own classrooms to their wider school community or indeed, across their provinces. Lonah and Jesse focused on developing assessment and curriculum knowledge and understanding within their school communities while Anna collaborated with colleagues to embed integrated STEM practices across various discipline areas and grades. Susan focused her attention on a province-wide professional learning initiative to educate her colleagues about STEM practices. Pauline's project was even more ambitious, on a national scale, where she specifically developed inclusive STEM curriculum practices for girls as an addition to the well-established Young Women Leaders (Yangpela Lida Meri) programme.

9.7.2 Participation in School Change

According to York-Barr and Duke (2004), teacher leaders are participants in school change through working with colleagues, being part of school wide decision making and being able to confront barriers to the changes being attempted. As already mentioned, the change taking place in our teacher leaders' narratives was the introduction of STEM in PNG schools as part of national reform. Lonah encouraged others to join her passion for STEM by having meetings about assessment practices. She then offered professional learning on assessment from which her colleagues joined her mission. Susan and Jesse had a similar approach, offering workshops from which their colleagues could learn more. Anna began by directly mentoring one beginning teacher who showed an interest in STEM pedagogies. Pauline had some ideas about people who shared her passion but spread the word more widely by developing a website. These teachers showed their leadership by working with school administrations to embed STEM professional learning programmes in their schools, districts, and nationally.

What is noteworthy, however, is that each of these teacher leaders has had to overcome and confront challenges for their initiatives to progress. Access to technology and connectivity were problems for all teacher leaders. Internet and electricity are often unreliable in PNG. Each teacher leader also encountered other problems, for example, Lonah and Pauline were hindered by lack of finances, and lack of time was a constraint for Lonah, Jesse and Susan. For Susan and Anna, getting their colleagues to share their passion and move their projects forward was sometimes difficult. Regardless, the teacher leaders persevered, understanding the wider implications of this reform in terms of social and economic benefits for their communities.

9.7.3 Coordination and Management

Curriculum change in schools needs coordination and management, tasks taken on by these five teacher leaders through the coordination of special events and participation in administrative tasks (York-Barr and Duke 2004). Lonah conducted meetings and interviews with Heads of Department and undertook an environmental scan of assessment practices in each department. She found many existing STEM practices being implemented unknowingly and used this information to coordinate a series of professional learning events. Susan and Jesse also coordinated professional learning for schoolteachers to introduce them to teaching and learning in STEM. Anna took advantage of school in-service sessions on the Standard-Based Curriculum to promote her integrated approach based around the school farm and her subject of Agriculture. Additionally, Susan was planning week-long intensive STEM teacher training programmes for the future and managing communication by creating a WhatsApp group for the professionally developed teachers to interact. She is currently in the process of coordinating a STEM fair where the "train the trainer" approach will be expanded to other schools in the district who are interested in STEM Education. Pauline has established advisory committees to develop a website where interested parties can access STEM resources specifically for girls.

9.7.4 Professional Development of Colleagues

Professional development of colleagues according to York-Barr and Duke (2004) includes leading workshops, modelling and encouraging professional growth. The professional development of colleagues to raise awareness and understanding of STEM has been the pivotal vehicle through which these teacher leaders have been able to lead positive change in their schools and communities. Anna's professional development of colleagues began with one teacher and has spread organically across the school. Lonah led in-service training with teachers across learning areas, in her school. The focus of this professional development was on instructional practices around planning learning and teaching programmes and assessments and rubrics associated with Project-based Learning in STEM. Jesse also focused on upskilling teachers in his own school, providing a series of 2-h professional development sessions on STEM teaching practices and pedagogies. While Susan also began her endeavours within her own school, initiating professional learning sessions for the Science teachers in her department, she also negotiated funds from her school principal to conduct two-day professional learning sessions on PBL and assessment for three secondary schools and five primary schools in her province. The purpose of these in-service training courses was to build teachers' capacities and encourage a collaborative, collegial community where practices could be shared. Her overall desired learning outcome was to improve teachers' STEM literacy. While Pauline did not lead any workshops, she encouraged the professional growth of many of her colleagues within and beyond her school.

9.7.5 Parent and Community Involvement

For York-Barr and Duke (2004), part of what teacher leaders do is encourage parent participation, create partnerships with community businesses and work with community and community organisations. Anna has encouraged the active participation of parents in sharing knowledge of Indigenous plants and animal husbandry. The other teacher leaders in this chapter have established other partnerships. For example, Lonah collaborated with the media, the Environment and Conservation Office, and the PNG Nature Park to conceptualise her Mini School Conservation Project. Susan's project was in collaboration with the Eastern Highlands Provincial Education Authority who embraced district-wide implementation of STEM. To expand her professional learning services, Susan also worked with the Director of Education for Adventist Agency Schools for expansion of STEM into primary education. For Pauline, she partnered with PNG Pass, Australian Department of Foreign Affairs, an association that partners Australian and PNG secondary schools. Jesse is the only teacher leader who did not write about partnering with any community organisation.

9.7.6 Contributions to the Profession

By contributions to the profession, York-Barr and Duke (2004) refer to participation in professional organisations, or in our cases, the creation of these and becoming politically involved. All five teacher leaders stepped outside their roles as classroom teachers and contributed to their profession through the creation of professional learning networks. For Lonah, Anna, Susan and Jesse, these networks consist of STEM teachers coming together to form communities of practice, where teachers are upskilled in STEM teaching and assessment practices, can share resources and have open communication about enablements and challenges. In Pauline's case, her network is all about young women and creating a STEM connection hub (website) for these girls to access STEM resources aimed at building confidence and changing mindsets about gender roles. These five teacher leaders have demonstrated political awareness as early adopters of the Standards-based Curriculum and inclusion of STEM into their practice. Their early uptake drove their desire to lead these curriculum initiatives locally, provincially and nationally.

So far, by analysing and synthesising these teacher narratives, we have a clear picture of "what teacher leaders do" to implement STEM Education and practices in PNG. However, what we don't know is "who these teacher leaders are" and the stance or dispositions they possess. In examining their stories, we found that the progression to being a teacher leader seems to be organic rather than imposed (Smulyan 2016). These individuals do not seem to be bound by their role rather what might have started as a set of behaviours and skills has morphed into deep ongoing commitment to collaborate, motivate and influence others towards a common vision (Collinson 2012). These teacher leaders have provided opportunities for conversation, learning and reflection (Smulyan 2016). They viewed each other as resources (Carver 2016) and encouraged others to take action. Their teacher leadership stances seem to be imbued with confidence, flexibility, care, a sense of community, creativity, inspiration, honesty, open communication, approachability, free exchange of ideas and empowerment. They went beyond the day-to-day work in the trenches, took risks, experimented, took a stand and spoke up (Rogers 2005). They exercised autonomy and professional judgement in a democratic sense of professionalism for their good, the good of their schools and colleagues and the community more broadly. Lowery-Moore et al. (2016) would refer to these stances or dispositions as these teacher leaders developing a new sense of emotional responsibility.

9.8 Final Thoughts

This chapter has presented the narratives of five teacher leaders as they embarked on implementing STEM practices in PNG. They and other teachers whose stories of leadership have been told in this book (see, e.g., Levesa Morumo 8.2 and Elias Wranga 8.4) have conveyed genuine convictions about striving for a better world, improving learning, teaching and assessing and creating communities of learning to translate ideas into sustainable actions. These teacher leaders became key change agents and sources of innovation (Nguyen et al. 2020) essential for building and expanding the STEM ecosystem (Yow et al. 2021) in PNG. They understood the importance of relationship building and collaboration to translate their vision into meaningful action and innovation. They used a range of strategies, including advocacy, agency, enabling others and risk-taking, to create a positive vision to bring about changes in their schools and communities.

By working as team players with a shared passion for making a difference, they re-cultured schools, built organisational capacity, and created lasting improvements in learning, teaching, and assessing. In other words, these teacher leaders epitomised instructional and distributed leadership at its best. Indeed, we would argue that these teacher leaders are also transformational leaders (Luyten and Bazo 2019) where practice was changed via professional learning communities and teacher learning. With the country's focus on economic growth and development, there is a growing need for a highly skilled workforce that can compete on the global stage. This requires a solid foundation in STEM Education, which can only be achieved through the collaborative efforts of dedicated teacher leaders.

References

Ainscow M, Sandill A (2010) Developing inclusive education systems: the role of organisational cultures and leadership. Int J Incl Educ 14(4):401–416. https://doi.org/10.1080/13603110802504903

Albion P (2015) Project-, problem-, and inquiry-based learning. In: Henderson M, Romeo G (eds) Teaching and digital technologies: big issues and critical questions. Cambridge University Press, pp 240–252

Allen M, Webb AW, Matthews CE (2016) Adaptive teaching in STEM: characteristics for effectiveness. Theory Pract 55(3):217–224. https://doi.org/10.1080/00405841.2016.1173994

Angelle P, DeHart C (2011) Teacher perceptions of teacher leadership: examining differences by experience, degree, and position. NASSP Bull 95(2):141–160. https://doi.org/10.1177/0192636511415397

Blackley S, Rahmawati Y, Fitriani E, Sheffield R, Koul RB. (2018). Using a makerspace approach to engage Indonesian primary students with STEM. Issues Educ Res 28(1):18–42. http://hdl.handle.net/20.500.11937/66478

Carver CL (2016) Transforming identities: the transition from teacher to leader during teacher leader preparation. Journal Res LeadShip Educ 11(2):158–180. https://doi.org/10.1177/1942775116658635

Chalmers C, Carter M, Cooper T, Nason R (2017) Implementing "big ideas" to advance the teaching and learning of Science, Technology, Engineering, and Mathematics (STEM). Int J Sci Math Educ 15(Suppl):25–43. https://doi.org/10.1007/s10763-017-9799-1

Chesky N, Goldstein R (2018) Packaging girls for STEM or STEM for girls? a critique on the perceived crisis of increasing female representation in STEM education. Crit Educ 9(16):98–126. https://doi.org/10.14288/ce.v9i16.186415

Collinson V (2012) Leading by learning, learning by leading. Prof Dev Educ 38(2):247–266. https://doi.org/10.1080/19415257.2012.657866

Curtis, R. (2013). Finding a new way: leveraging teacher leadership to meet unprecedented demands. Aspen Institute

Davila F, Bourke RM, McWilliam A, Crimp S, Robins L, van Wensveen M, Alders RG, Butler JRA (2021) COVID-19 and food systems in pacific island countries, Papua New Guinea, and Timor-Leste: opportunities for actions towards the sustainable development goals. Agric Syst 191(103137):1–11. https://doi.org/10.1016/j.agsy.2021.103137

DeMatthews D, Mawhinney H (2014) Social justice leadership and inclusion: Exploring challenges in an urban district struggling to address inequities. Educ Adm Q 50(5):844–881. https://doi.org/10.1177/0013161X13514440

Education Council (2018) Optimizing STEM industry-school partnerships: inspiring Australia's next generation. Final Report (Issue April). https://www.chiefscientist.gov.au/sites/default/files/2019-11/optimising_stem_industry-school_partnerships_-_final_report.pdf

Education Services Australia (n.d.) The girls in STEM toolkit (the GIST). https://www.esa.edu.au/solutions/our-solutions/girls-in-stem-toolkit

Femili PNG (n.d.) PSEAH (Preventing sexual exploitation, abuse and harassment). https://femilipng.org/about/complaints/pseah/

Forbes A, Chandra V, Pfeiffer L, Sheffield R (2021) STEM education in the primary school: a teacher's toolkit. Cambridge University Press

Gumus S, Bellibas MS, Esen M, Gumus E (2018) A systematic review of studies on leadership models in educational research from 1980 to 2014. Educational Management, Administration & Leadership 46(1):25–48. https://doi.org/10.1177/1741143216659296

Hmelo-Silver CE (2004) Problem-based learning: what and how do students learn? Educ Psychol Rev 16:235–266. https://doi.org/10.1023/B:EDPR.0000034022.16470.f3

Hunzicker J (2017) From teacher to teacher leader: a conceptual model. Int J Teach LeadShip 8 (2):1–.27. https://files.eric.ed.gov/fulltext/EJ1169800.pdf

Independent State of Papua New Guinea (2010) Papua New Guinea National Strategic Plan 2010–2050 "2050 Vision". https://png-data.sprep.org/system/files/2011.png.vision.2050.pdf

Ismail Z (2018) Benefits of STEM education. K4D helpdesk report. International development department. https://opendocs.ids.ac.uk/opendocs/handle/20.500.12413/14258

Jones S (2014) Distributed leadership: a critical analysis. Leadership 10(2):129–141. https://doi.org/10.1177/1742715011433525

Katzenmeyer M, Moller G (2009) Awakening the sleeping giant: helping teachers develop as leaders. Corwin Press

Ki-moon B (2014) Secretary-general's message for international women's day 2014. https://ourworld.unu.edu/en/secretary-generals-message-for-international-womens-day-2014#:~:text=Countries%20with%20more%20gender%20equality,anti%2Ddiscrimination%20and%20child%20support

Leithwood K (1999) Changing leadership for changing times. Int J Educ Manag 13(6):301–302. https://doi.org/10.1108/ijem.1999.13.6.301.4

Lertdechapat K, Faikhamta C (2021) Enhancing pedagogical content knowledge for STEM teaching of teacher candidates through lesson study. Int J Lesson & Learn Stud 10(4):331–347. https://doi.org/10.1108/IJLLS-03-2021-0020

Lowery-Moore H, Latimer RM, Villate VM (2016) The essence of teacher leadership: a phenomenological inquiry of professional growth. Int J Teach LeadShip 7(1):1–16. https://files.eric.ed.gov/fulltext/EJ1137503.pdf

Luyten H, Bazo M (2019) Transformational leadership, professional learning and learner centred teaching practices; evidence on their interrelations in Mozambican primary education. Stud Educ Eval 60:14–31. https://doi.org/10.1016/j.stueduc.2018.11.002

Mangin MM, Stoelinga SR (2008) Teacher leadership: What it is and why it matters. In: Mangin MM, Stoelinga SR (eds) Effective teacher leadership: using research to inform and reform. Teachers College Press, pp 1–9

Margolis J (2012) Hybrid teacher leaders and the new professional development ecology. Prof Dev Educ 38(2):291–315. https://doi.org/10.1080/19415257.2012.657874

Milgram D (2011) How to recruit women and girls to the Science, Technology, Engineering, and Math (STEM) classroom. Technology and Engineering Teacher 71(3):4–11

Muijs D, Harris A (2006) Teacher led school improvement: teacher leadership in the UK. Teach Teach Educ 22(8):961–972. https://doi.org/10.1016/j.tate.2006.04.010

Nguyen D, Harris A, Ng D (2020) A review of the empirical research on teacher leadership (2003–2017): evidence, patterns and implications. Journal of Educational Administration 58(1):60–80. https://doi.org/10.1108/JEA-02-2018-0023

Öqvist A, Malmström M (2016) Teachers' leadership: a maker or a breaker of students' educational motivation. School Leadership & Management 36(4):365–380. https://doi.org/10.1080/13632434.2016.1247039

Óskarsdóttir E, Donnelly V, Turner-Cmuchal M, Florian L (2020) Inclusive school leaders—their role in raising the achievement of all learners. Journal of Educational Administration 58(5):521–537. https://doi.org/10.1108/JEA-10-2019-0190

Ouedraogo I, Ayantunde AA (2020). ILRI Project Report: biomass production and management practices in mixed crop-livestock systems in the west African Sahel: opportunities and constraints. https://doi.org/10.13140/RG.2.2.36261.04329

PASS (PNGAus Partnership Secondary Schools). (n.d.). Ol Lida Meri: all women lead. https://www.pass.edu.au/professional-development/leadership/ol-lida-meri

Pedaste M, Mäeots M, Siiman LA, de Jong T, van Riesen S, Kamp ET, Manoli CC, Zacharia ZC, Tsourlidaki E (2015) Phases of inquiry-based learning: definitions and the inquiry cycle. Educ Res Rev 14:47–61. https://doi.org/10.1016/j.edurev.2015.02.003

Pineda-Báez C, Bauman C, Andrews D (2020) Empowering teacher leadership: a cross-country study. Int J Leadersh Educ 23(4):388–414. https://doi.org/10.1080/13603124.2018.1543804

PNG Education News (Papua New Guinea Education News) (2021, May 22) When and why was standard based education or curriculum introduced in PNG? https://edu.pngfacts.com/standard-based-education/when-and-why-was-standard-based-education-or-curriculum-introduced-in-png

PNGDOE (Papua New Guinea Department of Education) (2006) Agriculture. Subject field lower secondary syllabuses. https://educationpng.gov.pg/Teachers/secondary/lower/syllabus-lower-secondary-agriculture.pdf

PNGDOE (Papua New Guinea Department of Education) (2007) English lower secondary syllabus. https://educationpng.gov.pg/Teachers/secondary/lower/syllabus-lower-secondary-english.pdf

PNGDOE (Papua New Guinea Department of Education) (2008a) Business studies upper secondary syllabus. https://www.education.gov.pg/syllabus/secondarysyllabus/subjectssyllabus/Business%20Studies%20Grade%2011%20&12%20Syllabus.pdf

PNGDOE (Papua New Guinea Department of Education) (2008b) Biology. Upper secondary syllabus. https://www.education.gov.pg/syllabus/secondarysyllabus/subjectssyllabus/Biology%20Grade%2011%20&%2012%20Syllabus.pdf

PNGDOE (Papua New Guinea Department of Education) (2008c) Physics upper secondary syllabus. https://www.education.gov.pg/syllabus/secondarysyllabus/subjectssyllabus/Physics%20Grade%2011%20&%2012%20Syllabus.pdf

PNGDOE (Papua New Guinea Department of Education) (2023). Teachers. https://educationpng.gov.pg/Teachers/index.html

Poekert PE, Alexandrou A, Shannon D (2016) How teachers become leaders: An internationally validated theoretical model of teacher leadership development. Res Post-Compuls Educ 21:307–329

Portz S (2015) The challenges of STEM education. The space congress proceedings 43. https://commons.erau.edu/space-congress-proceedings/proceedings-2015-43rd/proceedings-2015-43rd/3/?utm_source=commons.erau.edu%2Fspace-congress-proceedings%2Fproceedings-2015-43rd%2Fproceedings-2015-43rd%2F3&utm_medium=PDF&utm_campaign=PDFCoverPages

Prinsley R, Beavis AS, Cliff-HordacreI N (2016) Busting myths about women in STEM. Office of the chief scientist, 13 (November), 1–4. https://www.uq.edu.au/about/files/6036/busting-myths-about-women-in-stem.pdf

Rogers J (2005) Aspiring to leadership-identifying teacher-leaders. Med Teach 27(7):629–633. https://doi.org/10.1080/01421590500156228

Sanga K, Reynolds M, Houma S, Maebuta J (2021) Tok stori as pedagogy: an approach to school leadership education in Solomon Islands. Aust J Indig Educ 50:377–384. https://doi.org/10.1017/jie.2020.31

Scoular C, Ramalingam D, Duckworth D, Heard J (2020) Assessment of general capabilities: skills for the 21st-century learner. Final report. Australian council for educational research. https://research.acer.edu.au/ar_misc/47

Sheffield R, Koul R, Blackley S, Maynard N (2017) Makerspace in STEM for girls: a physical space to develop twenty-first-century skills. Educ Media Int 54(2):148–164. https://doi.org/10.1080/09523987.2017.1362812

Silva DY, Gimbert B, Nolan J (2000) Sliding the doors: locking and unlocking possibilities for teacher leadership. Teach Coll Rec 102:779–804

Smulyan L (2016) Symposium introduction: stepping into their power—the development of a teacher leadership stance. Studies in Education 13(1):8–28. https://doi.org/10.1086/685800

Spillane JP (2011) The distributed leadership studies: a case study of research in and for school practice—research and practice pathways in mathematics education. In: Tate WF, Anderson CR, King K (eds) Disrupting tradition: research and practice pathways in mathematics education. National Council of Teachers of Mathematics, pp 7–19

Spillane JP, Halverson R, Diamond JB (2001) Investigating school leadership practice: a distributed perspective. Educ Res 30:23–28. https://doi.org/10.3102/0013189X030003023

Theoharis G (2007) Social justice educational leaders and resistance: toward a theory of social justice leadership. Educ Adm Q 43:221–258. https://doi.org/10.1177/0013161X06293717

Thornhill-Miller B, Camarda A, Mercier M, Burkhardt J-M, Morisseau T, Bourgeois-Bougrine S, Vinchon F, El Hayek S, Augereau-Landais M, Mourey F, Feybesse C, Sundquist D, Lubart T (2023) Creativity, critical thinking, communication, and collaboration: assessment, certification, and promotion of 21st century skills for the future of work and education. J Intell 11(3):54. https://doi.org/10.3390/jintelligence11030054

Torrance D (2015) Professional learning for distributed leadership: primary headteachers' perspectives. Prof Dev Educ 41(3):487–507. https://doi.org/10.1080/19415257.2014.936025

United Nations (UN) (n.d.) Sustainable development goals. https://www.un.org/sustainabledevelopment/sustainable-development-goals

Voigt-Graf C, Odhuno FO (2019) Assessing the labour market impact of the PNG LNG project and implications for future projects. Development Policy Centre Discussion Paper No. 78. https://doi.org/10.2139/ssrn.3360134

Waldron NL, McLeskey J, Redd L (2011) Setting the direction: the role of the principal in developing an effective, inclusive school. J Spec Educ LeadShip 24(2):51–60. https://doi.org/10.1177/0031721715569474

Wenner JA, Campbell T (2017) The theoretical and empirical basis of teacher leadership: a review of the literature. Rev Educ Res 87(1):134–171. https://doi.org/10.3102/0034654316653478

York-Barr J, Duke K (2004) What do we know about teacher leadership? findings from two decades of scholarship. Rev Educ Res 74(3):255–316. https://doi.org/10.3102/00346543074003255

Yow JA, Criswell BA, Lotter C, Smith WM, Rushton GT, Adams P, Ahrens S, Hutchinson A, Irdam G (2021) Program attributes for developing and supporting STEM teacher leaders. Int J LeadShip Educ 1–24. https://doi.org/10.1080/13603124.2021.2006794

Terri Bourke is a Professor at QUT. She has held a number of leadership positions including Course Co-ordinator, Academic Programme Director, Academic Lead Research and is currently the Dean and Head of School of Education. Her research interests include professional standards, professionalism, accreditation processes, assessment in geographical education and teaching about, to and for diversity. Her most recent research is related to teacher shortages. She, with Danielle Gordon, wrote the Introduction to and Review of this chapter *Teacher Leadership.*

Danielle Gordon is a Lecturer at QUT. She holds a Ph.D and teaches in multiple curriculum, pedagogy, and specialist studies in education units. Danielle's research interests include assessment and curriculum reform, teacher education and education psychology. Along with Terri Bourke, she has written the Introduction to and Review of this chapter Teacher Leadership.

Lonah Tombele with 19 years teaching experience, is the Head of Technology, Kilakila Secondary School, Port Moresby in the Sabama region. She is currently involved in teacher training and in-service development. Lonah has made contributions to Chap. 7 *Design Thinking in STEM* through a School Conservation Park Project and to this chapter *Teacher Leadership* outlining her approach to collaborative assessment in STEM.

Jesse Waison was, at time of writing, a teacher at Kerevat National High School, East New Britain with over 11 years of experience. He is now involved in the training of secondary Science and Mathematics teachers at St Peter Chanel Catholic College of Secondary Teacher Education. He has contributed to this chapter *Teacher Leadership* in introducing STEM and influencing STEM perspectives in a secondary school environment.

Anna Semese is an Agriculture and Biology teacher at Markham Valley Secondary School, Markham District, Morobe Province. She has over 10 years of teaching experience. She contributed to Chap. 8 *Project-based Learning* with a project to promote agriculture through access to sustainable crop irrigation.

Susan Totori was the Head of Science, Kabiufa Adventist Secondary School, Goroka at time of writing. She has since been promoted to the role of Deputy Principal (Academic) and has returned to the Queensland University of Technology to gain a Master of Education with a specialisation in STEM Education and Leadership. Further to this, she has been engaged as a part-time online tutor for the STEM Education studies for Science and Mathematics Postgraduate (PGDE) Teachers at the University of Goroka in the Eastern Highlands Province. She contributed to this chapter *Teacher Leadership* outlining distributed teacher professional learning in STEM.

Pauline Suvi is the Head of the Science Department and STEM Co-ordinator at the Sogeri National School of Excellence. At time of writing, she had 20 years of experience in teaching. She has recently established a consultancy dedicated to capacity building, professional development in STEM pedagogies, and international admissions and scholarships. She is planning to return to tertiary studies in STEM advocacy. Pauline contributed to this chapter *Teacher Leadership* concerning improving female participation in STEM through a digital platform.

10 The Way Forward for STEM Teacher Learning and Practice in Developing Countries

Vinesh Chandra, Margaret Lloyd, and Dann Mallet

This book has been about an Australian "STEM Education Teacher In-service Project" in PNG. It is a "complete book" which tells its story from multiple perspectives. It documents how a postgraduate course in integrated STEM Education was designed and delivered as well as presenting the voices of teachers who shared their experiences of professional learning and the challenges of implementation in the field. The teacher narratives highlight how their new knowledge and skills are impacting on their practices. Collectively, the book sheds light on an extraordinarily successful programme of teacher in-service development. Despite numerous unforeseen barriers (e.g., COVID-19 lockdown, poor connectivity), 96% of participants graduated with a Graduate Certificate in STEM Education, indicating their strong commitment and motivation to finish the course. By comparison, research suggests that generally 40% to 80% of students drop out of online classes, a figure that was even higher during the COVID-19 lockdown (Rahmani et al., 2024). The high success rate of the PNG teachers in this project is an indication that the Australian government funding expended was well spent and added significant value to the PNG education system.

V. Chandra (✉) · M. Lloyd · D. Mallet
Queensland University of Technology, Brisbane, QLD, Australia
e-mail: v.chandra@qut.edu.au

Oxfam (2017) reported that Australia currently gives about 22 cents for every $100 of its income in official development assistance, the lowest in Australian history. Chandra (2019) argued that "as funds become further depleted, the need for models for projects needs to change as well" (p. 66). As funding for Official Development Assistance (ODA) continues to fluctuate, this project could well provide a useful model for future investment. There were distinct characteristics of the project at the macro, meso, and micro levels (discussed in the sections to follow) which contributed to its success (see Fig. 10.1). It was evident that at each level, the stakeholders had a clear understanding of the "why", the "how", and the "what" questions (TEDx Talks, 2019).

10.1 Macrolevel

At the macrolevel were the needs of the country. In 2020, the scene was set with the introduction of the "Schools of Excellence Policy". The primary aspiration of the policy was to "produce students who are knowledgeable, highly educated, and agile in their abilities to comprehend scientific concepts and improvise to create innovative tools to solve everyday problems" (PNG PM Marape, 2020, para. 12). The introduction of the STEM curriculum for the National Schools of Excellence was a giant leap forward in terms of how

V. Chandra et al. (eds.), *Integrated STEM Education in Papua New Guinea*, Sustainable Development Goals Series, https://doi.org/10.1007/978-981-95-0828-0_10

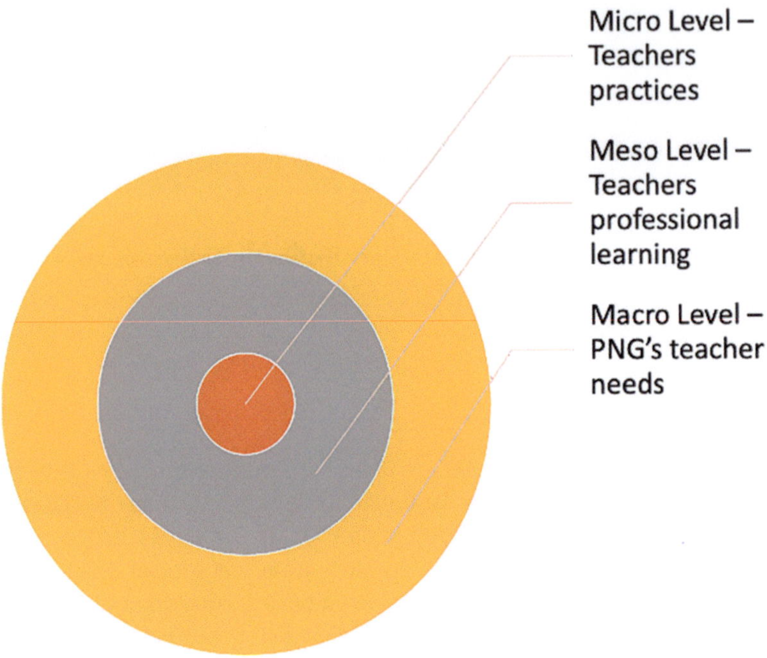

Fig. 10.1 The macro, meso and micro levels of the project

these subjects were to be delivered paving the way for schools to reimagine STEM Education. This vision was inspirational but also laid out aspirations for the country's education system. In this instance, it is evident that the PNG government had a clear policy of how it wanted its education system to be transformed.

Timely support by the Australian Government through its Official Development Assistance (ODA) programme set the wheels of teacher in-service in motion. Working collaboratively with the PNG government, Australia Awards PNG (AAPNG) was able to identify the type of professional learning programme that would best suit the teachers. Following the tendering process, our team was identified by Australia Awards as the successful bid to deliver a Graduate Certificate in STEM Education to support the Prime Minister's vision to transform PNG's education system.

From the feedback provided by the teachers, it was evident that they had all wanted to enrol in the course—they were not handpicked or forced into participating. They genuinely wanted to improve their pedagogical practices and knowledge of STEM Education in line with the government's vision. The "why" question was clear because teachers needed (and wanted) to engage in professional learning. The "how" question was clear with our university selected to deliver the postgraduate course due in part to a record of accomplishment in the delivery of postgraduate courses in integrated STEM Education. The "what" was noticeably clear in everyone's mind as a seamless integration of the STEM disciplines into teachers' daily practice in PNG schools.

10.2 Mesolevel

At the mesolevel was the design and delivery of the postgraduate certificate course. Understanding the needs of in-service teachers when designing such courses is particularly important. It helps us answer the "why" question. In this instance, the rationale of the course was to develop PNG teachers' pedagogical content knowledge (Shulman, 1987) in STEM Education. This assumed that in-service teachers have at least one subject specialisation with some teaching experience.

Our response to the "how" question is seen through the course design and delivery. We proposed and applied the "cyclical experiential learning model" based on the work of Bruner (1960) and Kolb (1984) because they progressively facilitate the development of teachers' knowledge and understanding together with their skills. We logically sequenced the subjects so that it enabled the teachers to develop their PCK in integrated STEM Education and then apply it in real-world contexts. Contextualising the course to the needs of the PNG teachers and their curriculum was a significant contributor to the success of this project. This approach supported the teachers' ability to further contextualise integrated STEM Education to their local contexts. Another feature of our design was that the subjects were offered in blocks which enabled the teachers to engage in a cycle of concrete experience, reflective observation, and abstract conceptualisation (Kolb, 1984) within and outside the course. The design of the course enabled the teachers to keep their focus on the key question: *How can I design and implement an integrated approach to teaching STEM in my context?*

The "what" question at the mesolevel asks, "what methods were used to deliver the course?" Our response was constrained by border restrictions. The COVID-19 pandemic rattled and challenged students as well as the teaching and management teams. Traditionally, Australia Awards short courses are delivered in a face-to-face/blended mode on an Australian university campus. Due to the pandemic, we pivoted to deliver the course through four different modes that included: aggregated on campus (intensive); aggregated off-campus (intensive); aggregated online (intensive); and, distributed online (short block). We were able to successfully deliver through all these modalities because all the parties associated with the delivery of the project worked in concert towards a common goal. DFAT, AAPNG, NDoE, QUT, school leaders, and awardees adapted to the meandering pathways presented by changing mandates enacted due to the COVID-19 virus to ensure that the course could be delivered with highest levels of effectiveness and efficiency under the prevailing circumstances. Consequently, we did not notice any significant differences in students' performances across the four modes. Feedback from the teachers demonstrated that all modes had their strengths and weaknesses some of which were based on personal preferences. Teacher in-service learning should incorporate mixed modal approaches to develop a variety of teaching and learning skills. However, for similar courses delivered in developing countries, teachers need to have access to appropriate digital technologies. In this project, the provision by AAPNG and DFAT of a laptop and high-speed Wi-Fi data access levelled the "digital" playing field. For a relatively small cost, it optimised teachers' participation in online learning and created pathways to ensure that no teacher was left behind.

10.3 Microlevel

The 18 unique teacher narratives presented in Part II of this book shed light on why, how and what questions at the micro level. All teachers were very clear on the "why" question. Each narrative had a rationale which outlined why the identified problem needed to be solved in response to teachers' contextual circumstances as well as the demands of their national curricula. The why question is often missed by teachers because they are too busy trying to meet the needs of the mandated curriculum so that it could be delivered in a timely manner. Quite often the question "why am I doing what I am doing?" is not asked. The teachers' narratives covered a broad range of topics and activities. Some were repeated indicating either common curriculum focuses or because of shared environmental issues that were repeated among the three cohorts of teachers. All, of course, were concerned with STEM (as an integrated whole or through component disciplines) and most framed their approach in collaboration with their teaching peers. The teachers' narratives describe STEM learning experiences which involve teacher professional learning, electricity, environmental care, water, heat, health, land use, earning an income, and food production. In addition, a small number of narratives were concerned with abstract rather than concrete goals: inequity, funding, and broadening student opportunities. Most were concerned with more than one topic e.g. concern with water supply was usually also associated with community health. The applications of STEM to address these issues is a visible extension of the SDGs being introduced as a framework for thinking.

The categorisation of the presented narratives as four discrete contexts of STEM Education: Authentic Learning, Design Thinking, Project-based Learning, and Teacher Leadership fitted together, and, interdependently, provided more credence to our response to the why question. Separately, each has a literature and exists outside of STEM. This is evident in the frameworks selected for analysis of the narratives.

- Authentic Learning—Herrington & Oliver's (2000) elements of AL framework
- Design Thinking—Henriksen et al.'s (2017) problem and solution definition
- Project-Based Learning—Krajcik & Shin (2022) features of PBL
- Teacher Leadership—York-Barr & Duke's (2004) domains of leadership

That these theoretical frames are used to analyse these narratives of STEM in PNG says something about the universality of the narratives and the level of abstract conceptualisation achieved by these teachers as they move through Kolb's (1984) experiential cycle.

The "how" of the narratives related to the pedagogies that the teacher authors adopted to follow in their implementation. Their choices were idiosyncratic and designed to suit their individual teaching styles and the resources available to them (see Project Summary Tables in each narrative). The range of selected pedagogies and their spread across the various aspects of STEM is shown in Table 10.1. Apart from Teacher Leadership, the pedagogies were deliberate choices of the teacher authors. Unsurprisingly, the STEM-PBL framework (Forbes et al., 2021) was adopted in three of the four Project-Based Learning narratives 8.1–8.3. The fourth PBL narrative chose the 5E Inquiry Framework (Bybee, 2010, 2015) as well as two others: the Entre Comp Framework (European Entrepreneurship Competence Framework (European Commission, 2018) and Visible Learning (Hattie & Yates, 2014). It is of interest, however, that STEM-PBL was also used in two Authentic Learning narratives (6.1–6.2). The remaining three Authentic Learning narratives respectively drew on Constructivism (6.3), 5E Inquiry (6.4), and Constructionism (6.5). There was considerable flexibility and personal professional judgement in how authentic learning experiences were structured. Similarly, design thinking was possible through different pedagogical approaches.

From our analysis, we contend that implementing integrated STEM within a context is not formulaic. There is no one "right" way and there can be productive overlaps and mashups between published frameworks and approaches. Pedagogy is dependent on context, resources, teacher choice, and course content. There is also variance in how teacher professional learning is structured and how teachers show leadership in this field of education especially in different cultural contexts. Teachers respond proactively to the circumstances they find themselves in.

The teachers' "what" questions were about the outcomes and their reflections, demonstrating what they had learnt. Each narrative demonstrated the way the teachers implemented their action plans and how they impacted their classroom practices. It enabled them to think more deeply about how they could design and deliver classroom activities where concepts in STEM could be connected to tackle problems that they or the students had identified themselves. Another significant component of the "what" question was the connection between what they were doing and the Sustainable Development Goals (see Appendix A). What they were doing in their classroom was contributing towards making the world a better place.

10.4 Final Words

This book has allowed a rare view of what STEM Education can provide for communities and whole economies. It shows that a concern for the greater good permeates teaching praxis. The book has also

Table 10.1 Pedagogies of the narratives

Pedagogy	Number of narratives	Authentic Learning	Design Thinking	Project-based Learning	Teacher Leadership
STEM-PBL Forbes et al. (2021)	5	6.1, 6.2		8.1–8.3	
5E Inquiry Learning Model Bybee (2010, 2015)	3	6.4	7.1	8.4	
Constructivism Amineh and Asl (2015)	1	6.3			
Social constructivism Amineh and Asl (2015)	1	6.3	7.2		
Constructionism Papert and Harel (1991)	2	6.3, 6.5			
d.school model of design thinking Hasso Plattner Institute of Design (2022)	3		7.3, 7.4		
Domains of leadership York-Barr and Duke (2004)	4				9.1–9.5
Entrecomp Framework European Commission (2018)	1			8.4	
Visible Learning Hattie and Yates (2014)	1			8.4	

presented a strategy on monitoring and evaluating the impact of an official development assistance (ODA) program that was funded by the Government of Australia and builds on the findings of the research undertaken by Hollingsworth et al. (2023). Four key strategies were proposed, two of which are particularly relevant to this book: first, investing in robust Monitoring and Evaluation (M&E) systems to track changes in teacher knowledge, attitudes, practices, and student learning, using this evidence to inform and refine professional development; and second, adopting "best fit" approaches that tailor programs to local contexts based on M&E insights. The teacher narratives presented in this book exemplify these strategies, highlighting how contextual adaptation and reflective practice empower teachers as leaders and experts in STEM education within their contexts.

Finally, this book has been written and compiled by a village. It is useful here to take the liberty of adding a "who" question to this circle because the people here, and the relationships between them, are most important. No one person—editor, teacher author or academic author—is more important than any other. Everyone has had a purpose in contributing to the whole and each is connected to each other through their praxis and in the abiding goodness of believing in the value of education. In this instance, it is education through integrated STEM which offers a more active and situated learning experience to students. We respectfully offer this experience to teachers and students in other learning contexts. There are lessons here which are as at home in a mountain village as they are in an urban environment.

References

Amineh RJ, Asl HD (2015) Review of constructivism and social constructivism. J Soc Sci, Lit Lang 1(1):9–16

Bruner J (1960) The process of education. Harvard University Press

Bybee RW (2010) Advancing STEM education: a 2020 vision. Technol Eng Teach 13(6):30–35

Bybee RW (2015) The BSCS 5E instructional model: creating teachable moments. NSTA Press

Chandra V (2019) Share engage educate—seeding change for a better world. Brill

European Commission (2018) Employment affairs and social inclusion. http://ec.europa.eu
Forbes A, Chandra V, Pfeiffer L, Sheffield R (2021) STEM education in the primary school: a teacher's toolkit. Cambridge University Press
Hasso Plattner Institute of Design (2022) d.school. Stanford University
Hattie J, Yates G (2014) Visible learning and the science of how we learn. Routledge
Henriksen D, Richardson C, Mehta R (2017) Design thinking: a creative approach to educational problems of practice. Think Ski Creat 26:140–153. https://doi.org/10.1016/j.tsc.2017.10.001
Herrington J, Oliver R (2000) An instructional design framework for authentic learning environments. Educ Technol Res Dev 48:23–48. https://doi.org/10.1007/BF02319856
Hollingsworth H, Wong D, Cassity E, Goundar P (2023) Teacher development multi-year study series: report highlights. Multi-country report. Aust Counc Educ Res. https://research.acer.edu.au/eas/72
Kolb DA (1984) Experiential learning: experience as the source of learning and development, Vol. 1. Prentice-Hall
Krajcik J, Shin N (2022) Project-based learning. In: Sawyer R (ed) The Cambridge handbook of the learning sciences (Cambridge Handbooks in Psychology). Cambridge University Press, pp 72–92. https://doi.org/10.1017/9781108888295.006
Oxfam (2017) Budget fails the fairness test. https://www.oxfam.org.au/2017/05/budget-2017-fails-the-fairness-test/
Papert S, Harel I (1991) Constructionism. Ablex
PNG PM Marape LAUNCHES SCHOOLS of Excellence Policy (2020, June 15) Papua New Guinea today. https://news.pngfacts.com/2020/06/png-pm-marape-launches-schools-of.html
Rahmani AM, Groot W, Rahmani H (2024) Dropout in online higher education: a systematic literature review. Int J Educ Technol High Educ 21:19. https://doi.org/10.1186/s41239-024-00450-9
Shulman LS (1987) Knowledge and teaching: foundations of the new reform. Harv Educ Rev 57:1–22. https://doi.org/10.17763/haer.57.1.j463w79r56455411
TEDx Talks (2019, November) How great leaders inspire action | Simon Sinek | TedxPugetSound. Ted Talk. https://www.ted.com/talks/simon_sinek_how_great_leaders_inspire_action/c
United Nations (n.d.) Sustainable development goals. https://www.un.org/sustainabledevelopment/sustainable-development-goals
York-Barr J, Duke K (2004) What do we know about teacher leadership? findings from two decades of scholarship. Rev Educ Res 74(3):255–316. https://doi.org/10.3102/00346543074003255

Vinesh Chandra is an Associate Professor and the STEM Subject Area Co-ordinator at Queensland University of Technology (QUT). In 2024, he was awarded an Australian Award for University Teaching (AAUT) for his contributions to advancing STEM education in Australia and overseas. He also served as one of the leaders for the *PNG STEM Education Teacher In-service Project.* For this project, the team was awarded a Gold in the QS Wharton Awards and a High Commendation in the IEAA awards. *In addition to being co-editor of this book, Associate Professor Chandra authored this chapter and contributed to others. He has also co-authored other books on STEM education, including the award-winning "STEM Education in the Primary School: A Teacher's Toolkit" published by Cambridge, and "My STEM Workbook" for primary school students.*

Margaret Lloyd is an Adjunct Professor, QUT, a Senior Fellow of the Higher Education Academy (HEA) and an Associate Fellow (Indigenous). She has won a national Teaching Excellence Award and is a National Teaching Fellow. She has had a long career in educational research and has worked extensively in academic professional development and international in-service teacher programs, particularly in S.E. Asia. Her editorial contribution to this book was mainly concerned with the design, collation and development of the teacher narratives presented and analysed in Part II: *Praxis—Teacher Narratives* and in grounding the teacher narratives within the Sustainable Development Goals.

Dann Mallet is Professor of STEM Education and Professor of Mathematics at Queensland University of Technology. He is experienced in academic and curriculum team leadership and mentoring, as well as extensive experience as a university Mathematics and STEM educator at all levels. Dann has initiated and led significant international education aid projects in the Indo-Pacific region, and has been recognised through numerous awards and nominations, including, most recently, a Gold in the QS Wharton Awards and High Commendation in the IEAA awards. Aside from his role as co-editor, he has contributed to Chaps. 3 and 4. Along with Vinesh Chandra, he wrote the Introduction to and Review of Chap. 8 *Project-Based Learning.*

Appendix A

Sustainable Development Goals

(United Nations (UN), n.d.).

SDG 1 ***No Poverty*** 	• **Authentic Learning in STEM** can help students understand the root causes of poverty and explore STEM strategies to address poverty at the local, national and global levels (Paula Niandros Polume 6.1) • **Project-based learning in STEM** affords investigation into activities which may generate income and financial stability in the future (Elias Wranga 8.4)

SDG 2 ***Zero Hunger*** 	• **Project-based learning** in STEM can be contextualised in agricultural and horticultural activities which may enable students to produce food for themselves and their communities at a time of increasing food insecurity in the developing world (Elias Wranga 8.4) • **Teacher leadership in STEM** encourages multidisciplinary activities which can sustain life through food production and security (Anna Semese 9.3)

V. Chandra et al. (eds.), *Integrated STEM Education in Papua New Guinea*, Sustainable Development Goals Series, https://doi.org/10.1007/978-981-95-0828-0

SDG 3 ***Good Health and Well-Being***	• **Authentic Learning in STEM** can encourage students to take an active role in ensuring their own health and well-being and that of their community (Debbie Zaccias 6.2; Anastasia Paraka 6.4; Koni Wamzil N'Drihin 6.5) • **Project-based Learning in STEM** can address specific aspects of health through improved nutrition (Paula Niandros Polume 8.1; Elias Wranga 8.4) and better living conditions (Abel P. Mambo 8.3) • **Teacher Leadership in STEM** can show students how to produce healthy food through integrated sustainable farming (Anna Semese 9.3)

SDG 4 ***Quality Education***	• **Authentic Learning in STEM** provides opportunities for students to engage in: hands-on activities where products embody learning (Paula Niandros Polume 6.1, Anastasia Paraka 6.4); experiential learning where students learn from real-world interactions (Gideon Aku 6.3); learning responds to real-world problems (Debbie Zaccias 6.2, Koni Wamzil N'Drihin 6.5) • **Design Thinking in STEM** offers the means for students to learn through applying and testing theory to practice (Douglas Hopkos 7.1, Lonah Tombele 7.2, Manasseh Kai Fena 7.3, Raphael Ulka 7.4.) • **Project-based Learning in STEM** offers a positive, productive and practical pedagogy for schools in PNG. It provides a platform for the integration of STEM knowledge and skills (Paula Niandros Polume 8.1; Levesa Morumo 8.2; Abel P. Mambo 8.3; Elias Wranga 8.4) • **Teacher Leadership in STEM** offers multiple pathways to support quality education by supporting teachers through in-service training, mentoring and sharing resources (Lonah Tombele 9.1; Jesse Waison 9.2; Anna Semese 9.3; Susan Totori 9.4; Pauline Suvi 9.5)

SDG 5 ***Gender Equality***	• **Teacher Leadership in STEM** actively supports the participation of girls and young women in STEM. Teachers work to overcome girls' lack of confidence and cultural stereotypes (Pauline Suvi 9.5)

SDG 6 ***Clean water and sanitation***	• **Design Thinking in STEM** can be used to produce collaborative solutions to ensuring a consistent and reliable water supply for a school and small community (Raphael Ulka 7.4) • **Project-based Learning in STEM** can be contextualised in improving community access to clean water and sanitation (Paula Niandros Polume 8.1)

SDG 7
Affordable and clean energy

- **Authentic Learning in STEM** allows students to focus on creating reliable sustainable and affordable forms of energy which make use of available resources in local communities (Debbie Zaccias 6.2; Koni Wamzil N'Drihin 6.5)
- **Design Thinking in STEM** provided a productive strategy to address the problem of an unreliable electricity supply in a remote school (Manasseh Kui Fena 7.3) and using natural elements to serve human needs (Douglas Hopkos 7.1)
- **Project-based Learning in STEM** can be contextualised in the generation of affordable and clean energy. This enables students to actively participate in an improved and sustainable living environment (Levesa Morumo 8.2; Abel P. Mambo 8.3)

SDG 8
Decent Work and Economic Growth

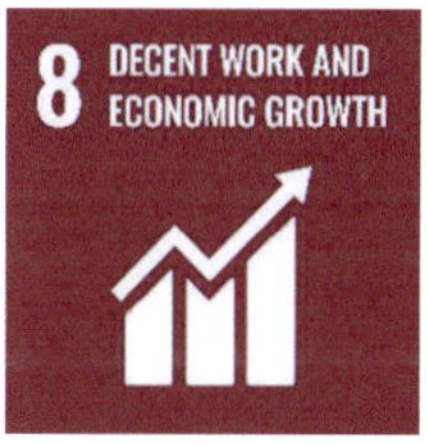

- **Authentic Learning in STEM** can help students to develop the skills and knowledge they need to succeed in the workforce and contribute to social and economic development. (e.g., production of recycled products) (Paula Niandros Polume 6.1; Anastasia Paraka 6.4)
- **Project-based Learning in STEM** can have a direct influence on decent work and economic growth in communities (Levesa Morumo 8.2; Elias Wranga 8.3)
- Teacher Leadership in STEM promotes the participation in STEM related careers (Jesse Waison 9.2, Pauline Suvi 9.5) and imagined futures in Agriculture (Anna Semese 9.3)

SDG 9
Industry, innovation and infrastructure

- **Authentic Learning in STEM** can afford students the opportunity to interact with industry (Gideon Aku 6.3)
- **Design Thinking in STEM** can afford students the opportunity to strategise and re-imagine water supply infrastructure (Raphael Ulka 7.4)
- **Project-based learning in STEM** can afford students the opportunity to better understand and propose improvements to power infrastructure (Levesa Morumo 8.2)

SDG 10
Reduced inequalities

- **Authentic Learning in STEM** can help in reducing inequalities through investigating ways to make daily life easier and more sustainable (Debbie Zaccias 6.2)
- **Design Thinking in STEM** can help in reducing inequalities by giving remote communities' access to electricity and consistent water supplies (Manasseh Kui Fena 7.3, Raphael Ulka 7.4)
- **Project-Based Learning in STEM** can help in reducing inequalities by experimenting with renewable energy (Levesa Morumo 8.2; Abel P. Mambo 8.3)
- **Teacher Leadership in STEM** can help in reducing inequalities by extending teacher professional learning across the provinces ensuring equitable delivery of new curriculum initiatives and sharing of resources (Susan Totori 9.4)

SDG 11
Sustainable cities and communities

- **Authentic Learning in STEM** can help students to develop an understanding of the challenges facing communities and the STEM tools and strategies needed to address these challenges (Paula Niandros Polume 6.1; Koni Wamzil N'Drihin 6.5)
- **Design thinking in STEM** can raise awareness and promote solutions to issues of waste and pollution (Manasseh Kui Fena 7.3)
- **Project-based Learning in STEM** can be situated in community settings to bring immediacy to learning (Elias Wranga 8.4)
- **Teacher Leadership in STEM** is critical to organising whole communities to work towards sustainable land use (Anna Semese 9.3)

SDG 12. Responsible Consumption and Production

- **Authentic Learning in STEM** can help students explore sustainable consumption and production patterns and develop STEM strategies to reduce waste and promote circular economies (Paula Niandros Polume 6.1; Debbie Zaccias 6.2; Anastasia Paraka 6.4; Koni Wamzii N'Drihin 6.5)
- **Design Thinking in STEM** can add to the challenge given to students in creating models and prototypes by restricting them to the use of waste or recycled materials (Douglas Hopkos 7.1, Lonah Tombele 7.2, Manasseh Kui Fena 7.3)
- **Teacher Leadership in STEM** offers students the opportunity to recycle or upcycle waste materials in their construction of prototypes (Anna Semese 9.3)

SDG 13
Climate Action

- **Design Thinking in STEM** can optimise learning experiences based on environmental projects and an awareness of climate change (Lonah Tombele 7.2)
- **Project-based learning in STEM** can raise awareness of climate action and focus on local small-scale contributions to a global problem (Abel P. Mambo 8.3)

SDG 14
Life below water

- **Authentic Learning in STEM** can support an awareness of the potential environmental damage caused by dumping plastic waste in the ocean (Paula Niandros Polume 6.1)

SDG 15 ***Life on land***	• **Design Thinking in STEM** can help students gain transferable skills in land and waste management, environmental regeneration and eliminating pollution to the soil and contaminates groundwater (Lonah Tombele 7.2; Manasseh Kui Fena 7.3) • **Teacher leadership in STEM** can involve students in sustainable integrated farming practices and growing Indigenous plants (Anna Semese 9.3) and environmental projects and based on land regeneration (Lonah Tombele 9.1)
SDG 16 **Peace, Justice and Strong Institutions**	• **Teacher Leadership in STEM**, through its establishment of self-sustaining, communities of practice and support networks create strong institutions which support teaching and learning (Susan Totori 9.4)
SDG 17 Partnerships for the goals	• **Design Thinking in STEM** can make use of an international partnership to improve opportunities for students and to provide support for hands-on projects. The PASS (PNGAus Secondary Schools) initiative aims to strengthen relationships between Papua New Guinean (PNG) and Australian secondary schools and to support STEM. (Raphael Ulka 7.4) • **Project-based Learning in STEM** can be promoted through partnerships with commercial and non-profit enterprises (Abel P. Mambo 8.3) • **Teacher Leadership in STEM** makes use of partnerships on a local and provincial level. (Lonah Tombele 9.1; Susan Totori 9.4; Pauline Suvi 9.5)

Appendix B
Mapping Narratives to PNG Curriculum Areas

Pedagogy	Narrative	Activity	SDGs (UN, n.d)
6. Authentic Learning	6.1	Weaving project	1, 4, 8, 11, 12, 14
	6.2	Sawdust stove	3, 4, 7, 10, 12
	6.3	Soap making, visits, model building	4, 9
	6.4	Soap making	3, 4, 8, 12
	6.5	Power generation	3, 4, 7, 11, 12
7. Design Thinking	7.1	Evaporation	4, 7, 12
	7.2	Conservation Park (Yrs 7-12)	4, 12, 14, 15
	7.3	Power generation	7, 10. 12, 15, 17
	7.4	Water supply	6, 9, 10, 17
8. Project-Based Learning	8.1	Water supply (Yr 9)	3, 4, 6
	8.2	Power generation (Yr 9 & 11)	4, 7, 8, 9, 10
	8.3	Power generation (Yr 11)	3, 4, 7, 10, 13, 17
	8.4	Entrepreneurship (Yrs 3, 4 & 5)	1, 2, 3, 4, 8, 11
9. Teacher Leadership	9.1	Assessment (Link to 7.2)	4, 15, 17
	9.2	Teacher PD (STEM Education)	4, 8, 9, 17
	9.3	School Farm	2, 3, 4, 8, 11, 12, 15
	9.4	Teacher PD (STEM Education)	4, 10, 16

V. Chandra et al. (eds.), *Integrated STEM Education in Papua New Guinea*, Sustainable Development
Goals Series, https://doi.org/10.1007/978-981-95-0828-0

	6 Authentic Learning					7 Design Thinking				8 Project-Based Learning				9 Teacher Leadership			
	1	2	3	4	5	1	2	3	4	1	2	3	4	1[11]	2[12]	3	4[3]
Agriculture							*							*	-	*	-
Applied Science		*					*	*	* 7			*		*	-		-
Biology			*				*							*	-		-
Bus. Stud.	*														-		-
Chemistry		*	*				*	*			*			*	-		-
D&T	*	*[4]	*	*[5]	*			*		*	*	*			-		-
Engineering		*										*			-		-
English							*							*	-		-
Geography							*							*	-		-
Maths (Jr)	*[1]		*	*						*	*		*[10]		-		-
Maths (Sr)		*[3]			*			*[6]			*[8]	*			-		-
Physics		*	*		*	*		*			*				-	*	-
Science (Jr)	*			*						*	*		*[9]		-		-

Notes to Table.

1. Measurement, Geometry (Polygons and Tessellations).
2. Principles and Practice (Craft Production).
3. Mathematics (Advanced): Number and Application: Ratio & Proportion, Units of Measurement, (Geometry) Congruency, similarity and construction
4. Design & Technology (Computer Studies, Metal Technology, Engineering).
5. Design & Technology (Computer Studies, Construction technology)
6. Advanced Maths
7. This was an extracurricular project, but the best fit is to Applied Science.
8. General Maths
9. Plant and Animal Science (Years 3, 4, & 5)
10. Maths (Years 3, 4, & 5)
11. Link to Narrative 7.2
12. Concerning PD on STEM principles
13. Concerning PD on SBC

Index

V. Chandra et al. (eds.), *Integrated STEM Education in Papua New Guinea*, Sustainable Development Goals Series, https://doi.org/10.1007/978-981-95-0828-0